Virginia's Continentals

1775-1778

~Volume One~

Michael Cecere

HERITAGE BOOKS
2022

HERITAGE BOOKS
AN IMPRINT OF HERITAGE BOOKS, INC.

Books, CDs, and more—Worldwide

For our listing of thousands of titles see our website
at
www.HeritageBooks.com

Published 2022 by
HERITAGE BOOKS, INC.
Publishing Division
5810 Ruatan Street
Berwyn Heights, Md. 20740

Copyright © 2022 Michael Cecere

All rights reserved. No part of this book may be reproduced or transmitted in any form or by any means, electronic or mechanical, including photocopying, recording or by any information storage and retrieval system without written permission from the author, except for the inclusion of brief quotations in a review.

International Standard Book Number
Paperbound: 978-0-7884-2486-1

Contents

Ch. 1　Virginia Raises Troops
　　　　April – October 1775 3

Ch. 2　Fighting Erupts in Virginia
　　　　September – December 1775 21

Ch. 3　The Virginia Continental Line Forms
　　　　December 1775-March 1776 45

Ch. 4　Preparation for War Increases
　　　　April-August 1776 71

Ch. 5　Virginia Regiments Join Washington
　　　　August -November 1776 97

Ch. 6　Trenton and Princeton
　　　　December 1776-January 1777 123

Ch. 7　Virginia's Continental Line Grows
　　　　Winter-Summer 1777 155

Ch. 8　Battle of Brandywine
　　　　September 1777 199

Ch. 9   Battle of Saratoga
                September-October 1777 223

Ch. 10   Germanton to Valley Forge
                October 1777-February 1778.................259

Appendix A: Virginia Continental Units
                1775-1778 299

Appendix B: Engagements of Virginia Continentals
                1775-1778 301

Appendix C: Organization of Virginia's
            Continental Line
                1777-1778 305

Bibliography 307

Index 319

Acknowledgements

As with all of my books, I'd like to thank my friends in the Revolutionary War reenacting hobby for their continued support and encouragement. I also want to thank the Rockefeller Library at Colonial Williamsburg and the research library at the Jamestown--Yorktown Foundation. These wonderful facilities provided me with the resources necessary to write this book. Thanks also goes to Debbie Riley of Heritage Books for her help with editing and another great book cover.

This book is dedicated to my dear friend, Evan "Buzz" Deemer. Buzz and I shared a passion for the American Revolution; and this book, like the others, is largely the result of our shared passion. I will always cherish his friendship and support.

Heritage Books by Michael Cecere:

*A Brave, Active, and Intrepid Soldier:
Lieutenant Colonel Richard Campbell
of the Virginia Continental Line*

*A Good and Valuable Officer:
Daniel Morgan in the Revolutionary War*

*A Universal Appearance of War:
The Revolutionary War in Virginia, 1775–1781*

*An Officer of Very Extraordinary Merit:
Charles Porterfield and the American War for Independence, 1775–1780*

Captain Thomas Posey and the 7th Virginia Regiment

*Cast Off the British Yoke:
The Old Dominion and American Independence, 1763–1776*

*Great Things are Expected from the Virginians:
Virginia in the American Revolution*

*He Fell a Cheerful Sacrifice to His Country's Glorious Cause:
General William Woodford of Virginia, Revolutionary War Patriot*

*In This Time of Extreme Danger:
Northern Virginia in the American Revolution*

*Second to No Man but the Commander in Chief:
Hugh Mercer, American Patriot*

*They Are Indeed a Very Useful Corps:
American Riflemen in the Revolutionary War*

*They Behaved Like Soldiers:
Captain John Chilton and the Third Virginia Regiment, 1775–1778*

*To Hazard Our Own Security:
Maine's Role in the American Revolution*

*Virginia's Continentals, 1775–1778
Volume One*

*Wedded to My Sword:
The Revolutionary War Service of Light Horse Harry Lee*

Introduction

When the Revolutionary War erupted in Massachusetts in April 1775, no American army existed. The only regular (full time) soldiers in the colonies were British redcoats, detachments of which had been posted in the colonies for decades. Twenty years earlier, many of the colonies, including Virginia, raised their own "regular" provincial soldiers to serve full time during the French and Indian War. Colonel George Washington commanded such troops for several years, but upon the end of that war, the colonists disbanded these units and returned to their reliance upon the British military and their own militia for their protection.

Each colony required its free male inhabitants, typically between the ages of 16 to 50, to serve in the militia, and such troops mustered for training several times a year. Few had seen any sort of conflict since the French and Indian War.

As tension grew between Britain and the colonies in the months prior to the outbreak of war at Lexington and Concord in April 1775, several colonies raised minute companies, select militia ready to respond at a moment's notice. Although such troops were likely superior than the ordinary militia, they were not full time, "regular" soldiers. The only such professional soldiers in the colonies remained British redcoats prior to April 1775. The Revolutionary War between Great Britain and the American colonies changed that.

This book explores the service of Virginia's continental troops during the first few years of the Revolutionary War. What began as two continental rifle companies sent north to Boston in the summer of 1775 became six regiments of continental troops by the end of the year and eventually sixteen continental regiments as well as two light dragoon regiments and a regiment of artillery. The service of all of these continental Virginians is the focus of this book.

Virginia in 1775

Chapter One:

Virginia Raises Troops

April – October 1775

In the weeks that followed the bloodshed on Lexington Green in Massachusetts on April 19, 1775, militia from nearly all of the thirteen American colonies took to the field in some form, some to organize and drill, others to threaten royal authorities, and still others to confront the British army in Boston. The militia that gathered outside of Boston in the spring of 1775 hailed from all of the New England colonies and formed a sort of ad hoc regional army of colonists. No continental authority had authorized the formation of these troops because the only continental authority that existed (outside of royal authority which was now being challenged) was the Continental Congress in Philadelphia, and it was initially unclear whether that body even had the authority, much less the will, to form a continental army.

The First Continental Congress had met in Philadelphia the previous fall of 1774 and debated for two months on how best to respond to the British parliament's crack down on Massachusetts (the Intolerable Acts). Although there was a brief discussion of improving the military preparedness of the colonies, most of the delegates at this first Congress rejected such talk as too provocative and instead, embraced economic sanctions to demonstrate their support for Massachusetts.[1]

---

[1] Paul H. Smith, ed., "Silas Deane's Diary, October 3, 1774," *Letters of Delegates to Congress: 1774-1789*, Vol. 1, (Washington, D.C.: Library of Congress, 1976), 138.

When the delegates to the Second Continental Congress gathered in Philadelphia in early May 1775, they were confronted with the issue of how to respond to the bloodshed in Massachusetts. Weeks of debate over the status of the troops in New England led to an agreement on June 14, 1775 to place them, as well as ten additional companies of riflemen from Pennsylvania, Maryland and Virginia, upon continental establishment.[2] George Washington, a Virginian, was selected by Congress to lead this new continental army, an army that now fell under the authority of all of the colonies through the Continental Congress.[3]

In Virginia, the counties of Berkeley and Frederick were tasked by Virginia's congressional delegation to raise the two rifle companies allotted to Virginia. The leaders of each county chose their rifle commanders well. Captain Hugh Stephenson was selected to lead the riflemen from Berkeley County while Captain Daniel Morgan was tapped to lead the Frederick County riflemen. Both men strove to recruit and equip their rifle companies as quickly as possible and then march north 600 miles to join their fellow Virginian, General Washington, in Massachusetts.

Henry Bedinger, a member of Captain Stephenson's company, recalled that,

> *Volunteers presented themselves from every direction, in the vicinity of these Towns;* [Winchester and Shepherdstown] *none were received but young men of*

---

[2] Worthington C. Ford, ed., "Proceedings of Congress, June 14, 1775," *Journals of the Continental Congress, 1774-1789*, Vol. 2, (Washington D.C.: U.S. Government Printing Office, 1905), 89.

[3] Ford, ed., Proceedings of Congress, June 15, 1775," *Journals of the Continental Congress, 1774-1789*, Vol. 2, 91.

Character, and of sufficient property to Clothe themselves completely, find their own arms, and accoutrements, that is, an approved Rifle, handsome shot pouch, and powder-horn, blanket, knapsack, with such decent clothing as should be prescribed, but which was at first ordered to be only a Hunting shirt and pantaloons, fringed on every edge, and in Various ways. Our Company was raised in less than a week. [Captain] *Morgan had equal success.*[4]

Peter Bruin, a rifleman in Captain Morgan's company, recalled that the challenge in raising the rifle companies was not in finding enough men, but choosing the best marksmen from all the volunteers.[5]

Captain Morgan and his company of nearly one hundred riflemen left Winchester on July 15th, crossing the Potomac River at Harper's Ferry.[6] When they reached the outskirts of Frederick, Maryland two days later, they were greeted by the local militia and escorted into town, "*amidst the acclamation of all the inhabitants that attended them.*"[7] One resident described their appearance:

Capt. Morgan, from Virginia, with his company of riflemen (all chosen), marched through this place on their way to Boston. Their appearance was truly

---

[4] Danske Dandridge, "Henry Bedinger to --- Findley," *Historic Shepherdstown*, (Charlottesville, VA: Michie Co., 1910), 79.
[5] John Dorman, ed., "Peter Bruin Application," *Virginia Revolutionary Pension Applications,* Vol. 12, (Washington, D.C., 1965), 3.
[6] Dandridge, "Henry Bedinger to --- Findley," *Historic Shepherdstown,* 81.
[7] Dandridge, "Extract of a Letter from a Gentleman in Frederick Town to His Friend, in Baltimore Town, Dated July 19, 1775," *Historic Shepherdstown*, 95.

> *martial; their spirits amazingly elated; breathing nothing but a desire to join the American army and to engage the enemies of American liberties....*[8]

The reception that Captain Morgan and his company received in Frederick was typical of nearly every community they marched through. All along the route the riflemen were greeted as heroes. Henry Bedinger, of Captain Stephenson's company (who trailed behind Morgan by several days) recalled that throughout the march,

> *We were Met by a Number of Men and Women out of the Country who Brought us churns of Beer, Cyder, and Buttermilk, apples, cheries, etc., etc. We honoured them by firing at our parting.*[9]

The arrival of the continental rifle companies in late July and early August sparked a wave of excitement among the New England troops outside Boston. Riflemen were largely unknown in New England and their appearance and reputation made quite an impression. Surgeon's Mate James Thacher of Massachusetts described the riflemen as

> *Remarkably stout and hardy men; many of them exceeding six feet in height. They are dressed in white frocks, or rifle-shirts, and round hats. These men are remarkable for the accuracy of their aim; striking a mark with great certainty at two hundred yards distance. At a review, a company of them, while on a quick advance, fired their balls into objects of seven inches diameter, at a*

---

[8] Ibid.
[9] Dandridge, "Henry Bedinger Journal," *Historic Shepherdstown*, 100.

distance of two hundred and fifty yards. They are now stationed on our lines, and their shot have frequently proved fatal to British officers and soldiers who expose themselves to view, even at more than double the distance of common musket-shot.[10]

Although the riflemen initially lived up to their reputation as deadly marksmen, the shortage of gunpowder in the American army soon compelled General Washington to curtail their frequent firing.[11] It seemed that whenever the riflemen engaged the enemy, scores of troops armed with muskets joined in, firing smoothbore weapons which were far less accurate than rifles. James Warren, President of the Massachusetts Provincial Congress, informed John Adams on August 9th, that

The General [Washington] *has been obliged from Principles of frugality to restrain his rifle men. While they were permitted Liberty to fire on the Enemy, a great number of the Army would go and fire away great quantitys of Ammunition to no Purpose.*[12]

In other words, to save gunpowder and shot, Washington reigned in the rifle companies, much to their chagrin.

Captain Morgan and his company were posted in the lines at Roxbury, directly across from the British fortifications on

---

[10] James Thacher, M.D., *Military Journal of the American Revolution*, (Gansevoort, New York: Corner House Historical Publications, 1998), 31.

[11] Philander D. Chase, ed., "General Orders, August 4, 1775," *The Papers of George Washington, Revolutionary War Series,* Vol. 1, (Charlottesville: University Press of Virginia, 1985), 218-19.

[12] Robert J. Taylor, ed., "James Warren to John Adams, August 9, 1775," T*he Papers of John Adams*, Vol. 3, (Cambridge, MA: Belnap Press, 1789), 114-115.

Boston Neck.[13] They, along with Captain Stephenson's rifle company and the two Maryland rifle companies, were part of the right wing of the American army outside of Boston.[14] As such, they fell under the command of General Artemas Ward. It was General Ward who relinquished command of the army to General Washington upon Washington's arrival in Cambridge in early July.

Restrained by General Washington's orders to limit gunfire from the lines, the riflemen chafed at the monotony of camp life and guard duty. Many grew restless and what little discipline existed among the riflemen deteriorated with the inactivity.

Captain Morgan and his men did not remain inactive for long, however. In early September, Morgan and two rifle company commanders from Pennsylvania, successfully drew lots to participate in an expedition through the Maine wilderness to capture the fortress city of Quebec.[15] The three rifle companies joined ten musket companies from the New England regiments, all under the command of Colonel Benedict Arnold of Connecticut, and together this 1,100 man force prepared for what would become an epic march through the wilderness of Maine to attack Quebec.

---

[13] William W. Williams, ed., "Robert Magaw to the Carlisle Committee of Correspondence, August 13, 1775," *Magazine of Western History*, Vol. 4, (May-October, 1886), 675.

[14] Ibid.

[15] Wm. H. Egle, ed., "The Journal of Captain William Hendricks and Captain John Chambers," *Pennsylvania Archives, 2nd Series*, Vol. 15, (1890), 31.

Virginia in 1775

Back in Virginia, the colony's rebellious leaders scrambled over the summer to address the alarming deficiency of their own military organization. An important militia law that had long been used to compel colonists to serve in the militia had inconveniently expired the previous year, throwing the whole militia system in Virginia into disarray. With the royal governor, John Murray, the Earl of Dunmore, self-exiled on a British warship in the York River out of fear for his safety, there was no chance that a new militia law would be approved to address the crisis.

Virginia's leaders had partially addressed this situation at the 2nd Virginia Convention in Richmond a month before Lexington and Concord and three months before Lord Dunmore fled Williamsburg. Patrick Henry declaration to, "Give Me Liberty, or Give Me Death," at the conclusion of his powerful address to the delegates in the Convention in late March, 1775,, highlighted Henry's argument that Virginia needed to put itself on a better defensive footing. In a close vote, the Convention agreed to encourage each county to form their own volunteer militia companies. Some of these militia volunteers, 250 strong, marched to Williamsburg in the summer of 1775 to protect the capital from feared attacks from Lord Dunmore and the British navy.[16]

---

[16] R.A. Brock, ed., "George Gilmer to Thomas Jefferson in Papers, Military and Political, 1775-1778 of George Gilmer, M.D. of Pen Park, Albemarle Co., VA," *Miscellaneous Papers 1672-1865 Now First Printed from the Manuscripts in the Virginia Historical Society,* (Richmond, VA, 1937), 101.

3rd Virginia Convention Authorizes Two Regiments of Regulars

Realizing by the summer of 1775 that these volunteer militia companies were inadequate for what was to come, the 3rd Virginia Convention disbanded the volunteers in August and replaced them with a three tiered military system that started with two regiments of provincial (regular / full time) soldiers raised for one year's service. The formation of two such regiments marked a significant escalation in Virginia's military preparedness.

After some debate and a close vote, Patrick Henry, of Hanover County, was selected by the 3rd Virginia Convention to command Virginia's 1st Provincial Regiment. It consisted of eight companies of 68 rank and file and totaled over 570 officers and men.[17] A number of Convention delegates resisted Henry's nomination citing his lack of any military experience, but his supporters overcame the opposition after several votes. Colonel William Woodford, of Caroline County (who did have experience as an officer in the French and Indian War) was given command of the second regiment. It comprised 476 rank and file in seven companies.[18]

To raise the troops for the two regiments, the Convention divided Virginia into sixteen districts and ordered each district (comprised of several counties) to recruit and send a company of regulars to Williamsburg as soon as possible.[19] The company of regulars from the eastern shore of Virginia

---

[17] William W. Hening, ed. *The Statutes at Large Being a Collection of all the Laws of Virginia,* Vol. 9, (Richmond: J. & G. Cochran, 1821), 9.
[18] Ibid.,10.
[19] Ibid., 10, 16.

remained there as a detached unit to help protect that isolated and vulnerable region of the colony.

The regular troops were not the only soldiers ordered to muster in Williamsburg; hundreds of minutemen were ordered to march to the capital as well. The minutemen comprised a second tier of Virginia's new military establishment. The Convention authorized sixteen battalions of minutemen. These men were drawn from the ranks of the militia and were *"more strictly trained to proper discipline"* than the ordinary militia.[20] Each of the sixteen districts was ordered to raise a 500 man battalion of minutemen *"from the age of sixteen to fifty, to be divided into ten companies of fifty men each."*[21] Like the regular troops, the minutemen were provided with proper arms as well as a hunting shirt and leggings.

The last tier of Virginia's new military establishment was the traditional county militia. The Convention decreed that

> *All free male persons, hired servants, and apprentices, above the age of sixteen, and under fifty years...shall be enlisted into the militia...and formed into companies....*[22]

As this provision pertained to Virginia's small free black male population, people of color served in Virginia's militia. They were not armed, however, and were used as drummers, pioneers, and other sorts of laborers.[23]

The Continental Congress in Philadelphia expanded the continental army, which originally consisted of the ten southern rifle companies and the troops in New England, to include

---

[20] Ibid., 16.
[21] Ibid., 16-17.
[22] Ibid., 27-28.
[23] Hening, ed., *Statutes of Law*, Vol. 7, 95 and Vol. 9, 268.

several New York regiments in the summer but that was the extent of the Continental Army entering the fall of 1775. Virginia's two regiments of regulars were thus not continental troops at their formation, but they were destined to be in just a few months.

Charged with defending the Old Dominion, the regimental officers of the 1st and 2nd Virginia Regiments received their provincial commissions in Newcastle from the Virginia Committee of Safety in mid-September.[24] Lieutenant Colonel William Christian of Fincastle County and Major Francis Epps of Dinwiddie County joined Colonel Patrick Henry in the 1st Virginia Regiment while Lieutenant Colonel Charles Scott of Cumberland County and Major Alexander Spotswood of Spotsylvania County served with Colonel William Woodford in the 2nd Virginia Regiment.

By the end of September, the first company of regulars under Captain John Markham from the Amelia District in central Virginia arrived in Williamsburg. Other companies soon followed. Lieutenant Colonel Thomas Bullett of Prince William County was appointed by the 3rd Virginia Convention to the post of adjutant general for the two regiments. It was his responsibility to organize the troops and their encampment in Williamsburg. Bullett's experience in the French and Indian War prepared him well for the task and he soon had a somewhat orderly encampment set up behind the College of William and Mary.

With both Colonel Henry and Colonel Woodford away from Williamsburg in early October, Lieutenant Colonel Bullett

---

[24] Robert L. Scribner and Brent Tarter, eds., "Proceedings of the Virginia Committee of Safety, September 19, 1775," *Revolutionary Virginia: The Road to Independence*, Vol. 4, (University Press of Virginia, 1978), 126.

initially commanded the growing garrison at the capital. Instructions on lodging, latrines, kitchens and victual [food provision] returns, as well as drill and daily camp routines such as reveille and retreat, troop returns, guard duty, and morning prayer service were all addressed by Lieutenant Colonel Bullett in his daily orders.[25] The first courts martial were held within a week of the encampment and by the time Colonel Henry arrived, on October 7th, the foundation for a well-disciplined military force had been established.[26]

That is not to suggest, however, that the regulars who arrived in the capital in the fall of 1775 were perfectly behaved, far from it. As a precaution against trouble, the troops were ordered to stay out of town unless they had permission from their company commanders, who in turn were ordered to allow no more than four men to be absent from their company at any time.[27] Camp guards were posted to enforce these orders.

Colonel Henry noticed a need to tighten discipline within days of his arrival and ordered that,

> *It is Expected that all Officers and Soldiers will preserve Decency and Good order in the Camp, and that Some irregularity now practiced be Reformed with as much haste as young Recruits will admit of.*[28]

---

[25] Brent Tarter, ed., "Orders for September 27 thru September 30, 1775," "The Orderly Book of the Second Virginia Regiment: September 27, 1775 – April 15, 1776," *The Virginia Magazine of History and Biography*, Vol. 85, No. 2 (April, 1977), 159-163.

[26] Tarter, ed., "Orders for September 30 and October 7, 1775," "The Orderly Book of the Second Virginia Regiment," *The Virginia Magazine of History and Biography*, Vol. 85, No. 2, 161-162.

[27] Tarter, ed., "Orders for September 28, 1775," "The Orderly Book of the Second Virginia Regiment," *The Virginia Magazine of History and Biography*, Vol. 85, No. 2, 161.

[28] Tarter, ed., "Orders for October 9, 1775" "The Orderly Book of the

Less than a week later, Colonel Henry had to remind the troops that they are, "*Again forbidden to Ease themselves, Except at the Necessary Holes....*"[29]

William Aylett, a prosperous merchant and planter from King William County who was appointed by the Virginia Committee of Safety to oversee the public storehouse in Williamsburg, undertook the monumental task of providing the troops with the fabric and equipment they needed to properly outfit and equip themselves.[30] Aylett's account books are full of disbursements of shoes, shirts, hats and buttons as well as a wide variety of fabrics (particularly oznaburg for hunting shirts), and a vast amount of camp equipment. When the men weren't on duty, they set to work sewing haversacks to carry their provisions, shot pouches to carry their musket and rifle balls, and hunting shirts, the designated uniform of the provincial regiments.[31]

The drill they practiced for several hours a day was His Majesty's 1764 Drill, the same exercise the British army employed and likely the same drill that most of the troops had learned as militia.[32] An additional drill, considered important to master but of which we know little detail about, was referred to

---

Second Virginia Regiment," *The Virginia Magazine of History and Biography*, Vol. 85, No. 2, 164.

[29] Tarter, ed., "Orders for October 15, 1775" "The Orderly Book of the Second Virginia Regiment," *The Virginia Magazine of History and Biography*, Vol. 85, No. 2, 168.

[30] Gregory B. Sandor, ed., *Journal of the Public Store at Williamsburg, 1775-1776*, (Columbus, Ohio, 2015) ii.

[31] Tarter, ed., "Orders for October 11, 1775" "The Orderly Book of the Second Virginia Regiment," *The Virginia Magazine of History and Biography*, Vol. 85, No. 2, 165-166.

[32] Tarter, ed. "Orders for October 12, 1775" "The Orderly Book of the Second Virginia Regiment," *The Virginia Magazine of History and Biography*, Vol. 85, No. 2, 166.

as the Woods Drill. Colonel Henry stressed the importance of this drill in his orders for October 10th.

> *The Officers and Soldiers to Spend one Hour to Day in Learning the Usual Exercise & Evolutions* [1764 Drill] *& are to Spend three Hours in Learning the Discipline of Woods fighting; to this last the Officers & men are to pay particular Attention and Compleat Themselves therein with all possible Deligence. Let it be Considered as the most Likely method to make the Troops formidable to the Enemies....*[33]

Men who perfected the assigned drill were placed in the grand squad where drilling continued but at a reduced intensity. Those who still needed more work were relegated to the awkward squad, where extra drill was conducted until they mastered the movements.[34]

Providing all of the troops with weapons and ammunition proved to be especially challenging. Colonel Henry initially wanted each soldier to have enough gunpowder and lead ball in their powder horns and shot pouches for ten shots, but that was reduced to six shots per man when it was discovered there was a severe shortage of both items in camp.[35]

---

[33] Tarter, ed., "Orders for October 10, 1775" "The Orderly Book of the Second Virginia Regiment," *The Virginia Magazine of History and Biography*, Vol. 85, No. 2, 165.

[34] Tarter, ed., "Orders for October 12, 1775" "The Orderly Book of the Second Virginia Regiment," *The Virginia Magazine of History and Biography*, Vol. 85, No. 2, 166-167.

[35] Tarter, ed., "Orders for October 13 and 15, 1775" "The Orderly Book of the Second Virginia Regiment," *The Virginia Magazine of History and Biography*, Vol. 85, No. 2, 167-168.

Based on the commission dates of the company captains of the two provincial regiments, it appears ten companies were in the capital by the middle of October.[36] The number of troops surpassed the space available in the college camp, so several companies were posted within the town itself.[37]

Colonel Woodford arrived sometime in the middle of October and on October 21st, Colonel Henry, the ranking officer among the two colonels, assigned the fifteen companies of regulars (some of who had not yet arrived in Williamsburg) to their respective regiments.[38]

Colonel Henry's 1st Virginia Regiment received Captain John Green's company from the Culpeper District (Culpeper, Orange and Fauquier Counties), Captain John Markham's company from the Amelia District (Amelia, Chesterfield and Cumberland Counties), Captain John Sayres's company from the Gloucester District (Gloucester, Essex, Middlesex, King and Queen and King William Counties), Captain William Davies's company from the Princess Anne District (Princess Anne, Norfolk, Isle of Wight, Nansemond Counties), Captain Robert Ballard's company from the Mecklenburg District (Mecklenburg, Charlotte, Halifax, Lunenburg, and Prince Edward Counties), Captain John Fleming's company from the Henrico District (Henrico, Goochland and Louisa Counties),

---

[36] E.M. Sanchez-Saavedra, ed., *A Guide to Virginia Military Organization in the American Revolution, 1774-1787*, (Westminster, MD: Willow Bend Books, 1978), 30-31, 36.

[37] Tarter, ed., "Orders for October 18 and October 20, 1775" "The Orderly Book of the Second Virginia Regiment," *The Virginia Magazine of History and Biography*, Vol. 85, No. 2, 168-169.

[38] Tarter, ed., "Orders for October 21, 1775" "The Orderly Book of the Second Virginia Regiment," *The Virginia Magazine of History and Biography*, Vol. 85, No. 2, 170-171.

Captain William Campbell's company from the Pittsylvania District (Pittsylvania, Botetourt, Bedford, and Fincastle Counties) and Captain George Gibson's company from West Augusta (which was essentially the Virginia frontier).[39]

Green's Campbell's and Gibson's companies were rifle companies and designated the regiment's light infantry. Although it is likely that a number of men in the remaining five companies also carried rifles, most carried smoothbore muskets and the companies were designated as line companies.

Colonel Woodford's 2nd Virginia Regiment consisted of Captain George Johnston's company from the Prince William District (Prince William, Fairfax, and Loudoun Counties), George Nicholas's company from the Elizabeth City District (Elizabeth City, Charles City, James City, New Kent, York, and Warwick Counties as well as Williamsburg), Captain Richard Parker's company from the Lancaster District (Lancaster, Westmoreland, Northumberland and Richmond Counties), Captain William Talaiferro's company from the Caroline District (Caroline, Stafford, King George, Spotsylvania Counties), Captain William Fontaine's company from the Buckingham District, (Buckingham, Amherst, Albemarle, and Augusta Counties), Captain Richard Kidder Meade's company from the Southampton District (Southampton, Sussex, Surry, Brunswick, Prince George and Dinwiddie Counties), and Captain Morgan Alexander's company from the Berkeley

---

[39] Tarter, ed., "Orders for October 21, 1775" "The Orderly Book of the Second Virginia Regiment," *The Virginia Magazine of History and Biography*, Vol. 85, No. 2, 170-171, and Sanchez-Saavedra, ed., *A Guide to Virginia Military Organizations in the American Revolution, 1774-1787*, 29-31.

District (Berkeley, Frederick, Dunmore and Hampshire Counties).[40]

Alexander's and Fontaine's companies were the two rifle companies for the 2nd Regiment, and like Captain Campbell's and Gibson's rifle companies in the 1st Virginia, had yet to arrive in Williamsburg. Captain John Green's rifle company was the sole rifle company among the two regiments then in Williamsburg and as a result, Green and his men found themselves very active in October and November.

---

[40] Tarter, ed., "Orders for October 21, 1775" "The Orderly Book of the Second Virginia Regiment," *The Virginia Magazine of History and Biography*, Vol. 85, No. 2, 171, and Sanchez-Saavedra, ed., *A Guide to Virginia Military Organizations in the American Revolution, 1774-1787*, 36.

Southeast Virginia

Chapter Two

Fighting Erupts in Virginia

September – December 1775

While Virginia's two provincial regiments of regulars scrambled to organize and equip themselves in Williamsburg, the Old Dominion's only continental soldiers, the two rifle companies of Captain Hugh Stephenson and Captain Daniel Morgan, found themselves in harm's way. Stephenson and his men remained in the lines outside of Boston with General Washington's continental army, occasionally skirmishing, but mostly enduring the standoff between the two sides.

Captain Daniel Morgan and his men however, boarded ships in Newburyport for a quick voyage to the Kennebec River in Maine (which was part of Massachusetts during the Revolution). Colonel Benedict Arnold commanded the 1,100 man expedition to Quebec, which arrived in detachments at Fort Western (in present day Augusta, Maine) in late September.

General Washington's plan was to threaten Quebec from two directions with two separate forces. While Arnold marched through the Maine wilderness to reach the British stronghold of Quebec, General Philip Schuyler would lead another American force into Canada via Lake Champlain. The expectation was that Schuyler's force would draw the British away from Quebec, leaving the fortified city vulnerable to attack by

Arnold.[1] Although this expectation did indeed have merit, other expectations, such as the distance of the route Arnold was to follow and the number of days it would take for his force to make it through the wilderness, were wildly off the mark.

Problems confronted the expedition almost immediately upon their departure from Fort Western in late September. As they made their way up the Kennebec River in hastily constructed bateaux made of unseasoned wood, the boats started to leak precipitously. The leaks grew worse as the heavily laden bateaux were battered by numerous shoals and rocks. Passage up the shallow river against a swift current proved extremely arduous for the troops, often requiring them to enter the cold river to guide the bateaux over difficult shoals. Several ledges and waterfalls also had to be overcome by grueling portages.[2]

Although two small scouting parties went ahead of the expedition in canoes, Captain Morgan's Virginians and the other two rifle companies from Pennsylvania made up the lead division of Arnold main body. They were tasked with blazing a trail for the rest of the expedition to follow across the Great Carrying Place, a twelve mile portage connecting the Kennebec River with the Dead River.

Before they even reached this spot, however, the expedition became aware of a crisis with their food supply. Water had seeped into many of the casks of food, spoiling much of the bread and peas, which had to be discarded. Dr. Isaac

---

[1] Chase, ed., "General Washington to Major General Philip Schuyler, August 20, 1775," *The Papers of George Washington, Revolutionary War Series,* Vol. 1, 332.

[2] Kenneth Roberts, ed., "Caleb Haskell's and George Morrison's Journal, September 28, 1775," *March to Quebec: Journals of the Members of Arnold's Expedition,* (New York: Country Life Press, 1938), 474, 511.

Senter recalled that, *"Our fare was now reduced to salt pork and flour...."*[3]

The situation grew worse when they reached the Great Carrying Place. Captain Morgan and the first division arrived on October 7th, and commenced to clear a trail for the detachments to come. Morgan, described by Pennsylvania rifleman John Joseph Henry as, *"large, [with] a commanding aspect and stentorian voice,"* worked alongside his men on the trail.[4] Dressed like his men, *"in the Indian style,"* (which included a fringed hunting shirt, breach cloth and leather leggings) and with his thighs all cut up by the brush and thorns, Morgan presented an imposing figure.[5]

Three ponds linked the long portage between the Kennebec and Dead Rivers, making the overland trek a bit easier. Nevertheless, hauling the heavy bateaux and supplies over the rough, steep, terrain between the ponds was a daunting task. Pennsylvania rifleman George Morrison described the experience.

> *This morning we hauled out our Batteaux from the river and carried thro' brush and mire, over hills and swamps (for we had not even the shape of a road but as we forced it) to a pond which we crossed, and encamped for the night. This transportation occupied us three whole days, during which time we advanced but five miles. This was by far the most fatiguing movement that had yet befell us, insomuch*

---

[3] Roberts, ed., "Isaac Senter's Journal, October 5, 1775," *March to Quebec: Journals of the Members of Arnold's Expedition*, 203.
[4] Roberts, ed., "John Joseph Henry's Journal, October 17, 1775," *March to Quebec: Journals of the Members of Arnold's Expedition*, 327.
[5] Ibid.

that we were often half leg deep in the mud, stumbling over old fallen logs, one leg sinking deeper in the mire than the other, then down goes a boat and the carriers with it, a hearty laugh prevails. The irritated carriers at length get to their feet with their boat, plastered with mud from neck to heel, their comrades tauntingly asking them how they liked their washing and lodging....[6]

Morrison continued:

Our encampments these two last nights were almost insupportable; for the ground was so soaked with rain that the driest situation we could find was too wet to lay upon any length of time; so that we got but little rest. Leaves to bed us could not be obtained and we amused ourselves around our fires most all the night...The incessant toil we experienced in ascending the river, as well as the still more fatiguing method of carrying our boats, laden with the provisions, camp equipage, etc., from place to place, might have subdued the resolution of men less patient and less persevering than we were.[7]

It was not possible to move everything across the portage in one haul, so the men trudged back and forth several times over the course of a week, sapping their strength, but not their determination to proceed.

---

[6] Roberts, ed., "George Morrison's Journal, October 9, 1775," *March to Quebec: Journals of the Members of Arnold's Expedition*, 513-514.
[7] Ibid., 514.

That determination was tested again in late October when the remnants of a hurricane swept in and flooded the Dead River, forcing the expedition to halt for four days and endure the pounding wind and rain.[8] Aware that his provisions were critically low and that a number of men were struggling to proceed, Colonel Arnold ordered the sick and those too weak to continue to return to Fort Western. He hoped that by doing so, the remainder of his force might squeeze two weeks of food out of the remaining provisions.[9]

The four divisions that made up Arnold's expedition had become strung out over many miles and Arnold was unaware of how demoralized the last division, under Lieutenant Colonel Roger Enos, had become. Thus, the news that Enos and his division had turned back came as a shock to Arnold in late October. Down more than a quarter of his men, Arnold pressed on, reaching the highest point of elevation on the journey, which required another grueling portage.[10]

While most of the bateaux had been abandoned by this point, Captain Morgan managed to maintain seven bateaux for his Virginians, but it came at a high cost. John Joseph Henry recalled that,

It would have made your heart ache to view the intolerable labors of these fine fellows. Some of them, it was said, had the flesh worn from their shoulders,

---

[8] Roberts, ed., "George Morrison's Journal, October 18-22, 1775," *March to Quebec: Journals of the Members of Arnold's Expedition*, 515.
[9] Roberts, ed., "Benedict Arnold's Journal, October 24, 1775," *March to Quebec: Journals of the Members of Arnold's Expedition*, 55.
[10] Stephen Clark, *Following Their Footsteps: A Travel Guide & History of the 1775 Secret Expedition to Capture Quebec.* (2003), 72-73.

even to the bone. By this time an antipathy had arisen against Morgan, as too strict a disciplinarian.[11]

The Virginians were not alone in their suffering, however. The great physical exertions of the last month and lack of provisions took a toll on all the men. George Morrison reported that

> *Never perhaps was there a more forlorn set of human beings...Every one of us shivering from head to foot, as hungry as wolves, and nothing to eat save a little flour we had left, which we made dough of and baked in the fires....*[12]

Dr, Isaac Senter confirmed the desperate situation the men faced in early November.

> *We had now arrived...to almost the zenith of distress. Several had been entirely destitute of either meat or bread for many days...The voracious disposition many of us had now arrived at, rendered almost anything admissible...In company was a poor dog* [who had] *hitherto lived through all the tribulations...This poor animal was instantly devoured, without leaving any vestige of the sacrifice. Nor did the shaving soap. Pomatum, and even the lip salve, leather of their shoes, cartridge boxes, Ect, share any better fate....*[13]

---

[11] Roberts, ed., "John Joseph Henry's Journal, October 28, 1775," *March to Quebec: Journals of the Members of Arnold's Expedition*, 335-336.

[12] Roberts, ed., "George Morrison's Journal, October 30, 1775," *March to Quebec: Journals of the Members of Arnold's Expedition*, 524.

[13] Roberts, ed., "Isaac Senter's Journal, November 1, 1775," *March to Quebec: Journals of the Members of Arnold's Expedition*, 218-219.

One last tribulation beset Captain Morgan and his company. While heading downriver in their bateaux, they came upon a set of rugged rapids that dashed their boats to pieces. Private Abner Stocking encountered Morgan's men a day after the incident and recorded in his journal that

> *We learnt to our great sorrow, that in attempting to go down the river in their batteaus...they were carried down by the rapidity of the stream and dashed on rocks; that they had lost most of their provisions and that a waiter of Captain Morgan was drowned. Their condition was truly deplorable – they had not...a mouthful of provisions of any kind, and we were not able to relieve them, as hunger stared us in the face.*[14]

Alas, relief finally reached the expedition when a small party of men that Colonel Arnold had sent ahead to find provision was spotted driving a number of livestock towards them. Arnold's expedition had made it through the wilderness. They had survived six weeks of hardship and misery on one of the most grueling and difficult marches in American military history. But there was still much to be done.

Fighting In Virginia

While Captain Morgan's company of Virginia riflemen struggled on the last leg of their march to Quebec, a company of Colonel William Woodford's 2nd Virginia Regiment kicked off armed hostilities in Virginia.

---

[14] Roberts, ed., "Abner Stocking's Journal, October 30, 1775," *March to Quebec: Journals of the Members of Arnold's Expedition*, 554-555.

Tensions between the royal governor, Lord Dunmore, and Virginia's leaders in Williamsburg had steadily escalated over the summer and into the fall of 1775. Dunmore, safe aboard a British warship in the York River, sailed to Norfolk in July and welcomed two small detachments of British regulars from the 14th Regiment posted in St. Augustine in August and October.[15] Emboldened by these reinforcements as well as the presence of British warships *HMS Mercury* (20 guns, 170 men) and *HMS Otter,* (16 guns, 100 men) Dunmore launched a series of raids in October to seize weapons and gunpowder in Norfolk and Princess Anne County against no opposition.

Captain Matthew Squire of the *Otter* did not participate in these raids. He was focused on striking at the inhabitants of Hampton, across Hampton Roads from Norfolk, to retaliate for an incident in early September that cost him a tender (small support vessel) and embarrassed him dearly.

In early September, Captain Squire, who was aboard a tender, was swept up by a hurricane that struck Hampton Roads, driving his small vessel ashore near Hampton. The local militia, angered by recent British naval raids upon the shore for livestock and provisions, seized the opportunity for payback by capturing the vessel and most of its crew. Captain Squire managed to escape, however, and eventually made his way back to the *Otter,* where he demanded the return of the tender and its stores.[16]

---

[15] Alexander Purdie, "August 4, 1775," *Virginia Gazette*, 3, and Peter Force, ed., "Monthly Return of His Majesty's Forces in the Province of East Florida, September and October, 1775," *American Archives, Fourth Series*, Vol. 4, (Washington, DC: U.S. Congress, 1848-1853), 323-326.

[16] William B. Clark, ed., "Captain Squire to the Hampton Town Committee, September 10, 1775," *Naval Documents of the American Revolution*, Vol. 2, (Washington, 1966), 74.

Hampton's leaders brazenly denied any involvement in the incident and appealed to the Committee of Safety in Williamsburg for help. They sent a hundred volunteer militia, the remnants of the volunteers who had protected Williamsburg over the summer.[17]

By October, these volunteers were replaced by a company of regulars from the 2nd Virginia Regiment under Captain George Nicholas, the son of Virginia's treasurer, Robert Carter Nicholas. Captain Nicholas's company was raised in the area and a good choice to help defend Hampton. They were joined by a company of minutemen as well as the local militia.

Captain Squire had remained aboard the *Otter* in Hampton Roads since the loss of his tender, effectively blockading Hampton's harbor. On October 26th, 1775, he made his move, leading a small squadron of support vessels consisting of a large schooner, two sloops, and two pilot boats, into the Hampton River. His intention was to sail up to Hampton, less than a mile upriver, to bombard and burn the town. His passage was blocked, however, by several sunken vessels obstructing the channel.[18]

Lord Dunmore, who did not participate in this engagement but who likely received a report of it from Captain Squire, described what happened in a report to London.

> *Some of the King's tenders went pretty close into Hampton Road. So soon as the rebels perceived them, they marched out against them and the moment they got within shot of our people, Mr. George*

---

[17] Scribner and Tarter, eds. *Revolutionary Virginia: The Road to Independence*, Vol. 4, 96.
[18] John Dixon and Wm. Hunter, "October 28, 1775," *Virginia Gazette*, 3.

> *Nicholas…who commanded a party of rebels at that time in Hampton, fired at one of the tenders, whose example was followed by his whole party. The tenders returned the fire but without the least effect.*[19]

Another account credited the commander of the minute company, Captain George Lynn with firing the first shot while others blamed it on Squire's tenders.

While it remains unclear who fired the first shot at Hampton, the result was the commencement of war in the Old Dominion in the form of a pitched ship to shore battle that lasted over an hour.[20] Unable to maneuver past the sunken vessels obstructing the harbor, Captain Squire's stationary tenders were raked with rifle and musket fire from shore. Their crews responded with cannon, swivel, and musket fire, but it had little effect on the well protected Virginians on shore. One rebel combatant recalled that

> *The fire* [from the tenders] *consisted of 4 pounders, grape shot etc. for about an hour. Not a man of our's was hurt. Whether our men did any damage is uncertain. They could not get nigher than 300 yards. Some say they saw men fall in one of the tenders.*[21]

The fighting subsided when the British withdrew out of range, but they returned during the night and worked in a driving rain

---

[19] Davies, K.G. ed., "Lord Dunmore to Lord Dartmouth, December 6, 1775 through February 18, 1776," *Documents of the American Revolution*, Vol. 3, (Shannon: Irish University Press, 1972), 58.

[20] Pinkney, "November 2, 1775," *Virginia Gazette*, 3.

[21] Ibid.

to clear the obstructed channel so that they could sail on to Hampton.

While the British worked to free the passage, Colonel Woodford, accompanied by a company of riflemen from the Culpeper Minute Battalion, marched all night from Williamsburg to reinforce Hampton's defenders and take command of the situation. John Page, a member of the Committee of Safety, described to Thomas Jefferson, what occurred. The detail of Page's account suggests that he likely obtained the information from Colonel Woodford himself.

Col. Woodford accompanied Captain [Abraham] Buford's rifle company through a heavy rain to Hampton and arrived about 7 a.m. When the Col. Entered the Town, having left the Rifle Men in the Church to dry themselves, he rode down to the River, took A view of the Town, and then seeing the Six Tenders at Anchor in the River went to Col. Cary's to dry himself and eat his Breakfast. But before he could do either the Tenders had cut their Way through the Vessel's Boltsprit which was sunk to impede their Passage and having a very fresh and fair Gale had anchored in the Creek and abreast of the Town. The People were so astonished at their unexpected and sudden Arrival that they stood staring at them and omitted to give the Col. the least Notice of their approach. The first intelligence he had of this Affair was from the Discharge of a 4 Pounder. He mounted his Horse and riding down to the Wharf found that the People of the Town had abandoned their Houses and...the Militia had left the Breast Work which had been thrown up across the Wharf and street.

He returned to order down Captn. Nicholas's Company and Buford's and meeting Nicholas's which had been encamped near Col. Cary's he lead them pulling down the Garden pails [fence] through Jones's Garden under Cover of his House, and lodged them in the House directing them to fire from the Window which they did with great Spirit. He then returned and led Buford's Company in the same manner under Cover of Houses and others at a Breast work on the Shore.... Captn Barron with the Town Militia and Part of Nicholas's Company were stationed at the Brest Work on the Wharf and across the Street. The fire was now general and constant on both Sides. Cannon Balls, Grape Shot and Musket Balls whistled over the Heads of our Men, Whilst our Muskets and Rifles poured Showers of Balls into their Vessels and they were so well directed that the Men on Board the Schooner in which Captain Squires himself commanded were unable to stand to their 4 Pounders...but kept up an incessant firing of smaller Guns and swivels, and 3 Boats for more than an Hour....[22]

The intensity of the rebel small arms fire from shore, particularly from the Culpeper riflemen that had arrived with Woodford, eventually forced the British ships to withdraw with the loss of several men and a small tender.[23] The several Virginia gazettes celebrated the brave stand of Hampton's defenders, one proclaiming that, *"Lord Dunmore may now see he has not cowards to deal with."*[24]

---

[22] Clark, ed., "John Page to Thomas Jefferson, November 11, 1775," *Naval Documents of the American Revolution*, Vol. 2, 991-992.

[23] Dixon and Hunter, "October 28, 1775," *Virginia Gazette*, 2.

[24] Pinkney, "November 2, 1775," *Virginia Gazette*, 3.

With Hampton secure, Colonel Woodford returned to Williamsburg with Captain Nicholas's company to execute yet another order from the Committee of Safety. Woodford was to lead his regiment, supported by five companies of Culpeper minutemen, across the James River to march to Norfolk and confront Lord Dunmore's small force gathered there.

The decision of the Committee of Safety to send Woodford and the 2nd Regiment instead of Colonel Patrick Henry and the 1st Regiment did not sit well with Henry and his supporters. As the commander-in-chief of Virginia's forces, he believed the honor of leading the troops to Norfolk belonged to him. The Committee of Safety, wary of Henry's lack of military experience, thought otherwise and ordered Woodford to take his regiment, instead. Their departure was delayed, however, by a shortage of supplies and equipment and the presence of the British warship *HMS King Fisher* (20 guns and 170 men) with several tenders off of Jamestown Island. They were positioned to block any passage across the river from the ferries at Burwell Landing or Jamestown Island.

Captain John Green's rifle company of the 1st Virginia was still the only company of regular rifle troops in Williamsburg in early November, so the job of confronting the British ships fell to them. Their rifles, with their extended accuracy (compared to muskets) proved to be more than an adequate deterrence to the British ships, who responded with ineffective swivel and cannon shot.[25] John Page noted the effectiveness of the riflemen in a letter to Thomas Jefferson.

---

[25] Purdie, "November 10, 1775," *Virginia Gazette*, 3.

> *I can assure you that about 20 Rifle Men have disputed with the Man of War and her Tenders for…2 Days and they have hitherto kept…the Ferry Boats safe, which it is supposed they wish to burn. It is incredible how much they dread a Rifle.*"[26]

Edmund Pendleton, the president of the Committee of Safety, also acknowledged the role of Captain Green's riflemen in defending the shoreline near Williamsburg.

> *The life and Soul of* [the troops guarding the shore] *is Capt. Green's Company of Riflemen from Culpeper, who in three Reliefs of about 22 at a time, scour the River, and have in various Attempts, prevented a landing of the enemy. Last week the King Fisher and four tenders full of men came up to Burwell's Ferry and made several attempts to land during three days stay, but never came nearer than to receive a discharge of the Rifles, when they retired with great precipitation, and 'tis Supposed the loss of some men.*[27]

Although Captain Green's riflemen were able to prevent the British from landing onshore, they could not force them from the river, so Colonel Woodford and his force of approximately 660 troops had to march upriver to Sandy Point, where the James River narrowed considerably, and

---

[26] Clark, ed., "John Page to Thomas Jefferson, November 11, 1775," *Naval Documents of the American Revolution*, Vol. 2, 991-992.

[27] David John Mays, ed., "Edmund Pendleton to Thomas Jefferson, November 16, 1775," *The Letters and Papers of Edmund Pendleton*, Vol. 1, (Charlottesville: The University Press of Virginia, 1967), 130.

cross there.[28]   They did so in mid-November and then proceeded south, towards Suffolk and Great Bridge, where Governor Dunmore had erected a wooden fort to prevent the rebels from marching upon Norfolk itself.

Battle of Great Bridge

While Colonel Woodford made his way towards Great Bridge, Governor Dunmore enjoyed much success in Norfolk, routing a detachment of Princess Anne County militia at Kemp's Landing in mid-November and then welcoming the loyalty oaths of hundreds of inhabitants responding to Dunmore's proclamation calling on all Virginians to rally to the King's standard, *"or be looked upon as Traitors to His Majesty's Crown and Government."*[29] Dunmore also offered freedom to all runaway slaves and indentured servants of rebels who agreed to fight for him.[30]

He gleefully reported to General William Howe in Boston, the new British commander in North America, that his proclamation, *"has had a Wonderful effect as there are not less than three thousand* [people] *that have already taken and signed the* [loyalty] *Oath."*[31] He also noted that, *"the Negroes are flocking in also from all quarters which I hope will oblige the Rebels to disperse to take care of their families and property."* Dunmore added that, *"had I but a few more men*

---

[28] Ibid.
[29] Clark, ed., "Lord Dunmore's Proclamation, November 14, 1775," *Naval Documents of the American Revolution*, Vol. 2, 920.
[30] Ibid.
[31] Clark, ed., "Lord Dunmore to General William Howe, November 30, 1775," *Naval Documents of the American Revolution*, Vol. 2, 1209-1211.

here I would March immediately to Williamsburg my former place of residence by which I should soon compel the whole Colony to submit".[32]

Although his situation looked promising in November, Dunmore lacked muskets to supply his growing number of volunteers and his detachment of British regulars from the 14th Regiment only numbered approximately 150 officers and men.[33] He was fortunate, however, to have control of both the water and land passages to Norfolk. Several British warships secured the Elizabeth River, while outposts at Kemp's Landing and Great Bride secured the land routes to Norfolk.

The Great Bridge, eleven miles south of Norfolk, was actually a long, narrow, manmade causeway with multiple wooden bridges spanning the southern branch of the Elizabeth River and its tributaries and marshland. This chokepoint was the key to holding Norfolk, at least until earthworks were completed on the southeast side of the town.

Recognizing the importance of holding the Great Bridge, Dunmore erected a stockade fort on the north side of the dismantled bridge and garrisoned it with a detachment of British regulars supported by a number of Tory and black soldiers.

It was this force that Lieutenant Colonel Charles Scott of the 2nd Virginia Regiment first encountered when he arrived at Great Bridge in late November with the advance guard of Colonel Woodford's force. Some skirmishing erupted with little loss to either side and Scott erected a strong breastwork

---

[32] Ibid.
[33] Peter Force, ed., "Monthly Return of His Majesty's Forces in the Province of East-Florida, October 1, 1775," *American Archives, Fourth Series*, Vol. 4, (Washington, 1837), 325-326.

across the causeway, about three hundred yards from the fort, to protect his troops from a sudden attack from the fort. A weeklong standoff ensued during which several sharp skirmishes occurred downriver and there was daily firing between the two sides at Great Bridge, but it had little effect on the stalemate.

Colonel Woodford and the bulk of his force encamped several hundred yards south of the breastworks, within the village of Great Bridge, when they arrived in early December. He updated the Committee of Safety:

> *I...found the Enemy Posted on the opposite side of the Bridge, in a Stockade Fort, with two four pounders, some swivels & wall pieces, with which they keep up a constant Fire....We have raised a strong Breastwork upon the lower part of the street joining the Causeway, from which Centries are posted at some old Rubbish not far from the Bridge (which is mostly destoy'd).*[34]

Although the shortage of gunpowder forced Colonel Woodford to restrict his men from returning fire upon Dunmore's fort, he did post sentries between the fort and breastworks in the evening to serve as pickets against a possible surprise attack from the British. They were withdrawn back to the breastworks at sunrise each day.

Colonel Woodford appealed to Virginia's leaders for more gunpowder as well as additional blankets for his men.

---

[34] D.R. Anderson, ed., "Colonel Woodford to Edmund Pendleton, December 4, 1775," *Richmond College Historical Papers*, (June, 1915), 106-107.

The men are tolerably well...but the dampness of this Ground, without straw (which is not to be had) must soon lay many of them up, & Houses that are tolerably safe from the Enemy's Cannon, can only be procured for a few."[35]

Help was on the way for Colonel Woodford and his troops in the form of reinforcements from North Carolina who reportedly were bringing several cannon with them. It was the prospect of cannon that caused Governor Dunmore to make what proved to be a rash decision, attack the rebels before the reinforcements arrived.

His plan called for two companies of black troops to cross a few miles downriver and make their way to attack the rear of the rebel encampment. When this diversion commenced, Captain Samuel Leslie with the bulk of the British regulars, would march out of the fort, supported by more black and Tory troops, and directly assault the rebel breastworks.[36]

The expectation was that with the main rebel encampment, distracted by the two companies in their rear, the rebels posted at the breastworks would easily fall to the British regulars. Dunmore's plan fell apart, however, when the diversionary force failed to even cross the river because of a misunderstanding of orders. Captain Leslie, with over 350 men, unwisely proceeded with the attack, marching out of the fort at dawn.[37]

---

[35] Ibid.
[36] William B Clark, ed., "Lord Dunmore to Lord Dartmouth, December 6, 1775 through February 18, 1776," *Naval Documents of the American Revolution*, Vol. 3, (Washington, 1968), 141.
[37] Clark, ed., "Letter to John Pinkney, December 20, 1775," *Naval Documents of the American Revolution*, Vol. 3, 186-189.

If the handful of sleepy rebel sentinels posted behind piles of debris between the fort and rebel breastworks failed to notice British efforts to restore the bridge as the sun rose on December 9th, the discharge of the fort's cannon likely drew their attention that way. Startled at what they saw, the sentries opened fire, most firing several rounds and one, Billy Flora, a free black militiaman, firing up to eight rounds before they scurried back to the breastworks 200 yards away.[38] Although the sixty or so men behind the rebel breastworks realized that an attack was coming, the bulk of the rebel camp initially remained unaware. Major Alexander Spotswood of the 2nd Virginia recalled that,

We were alarmed this morning by the firing of some guns after reveille beating, which, as the enemy had paid us this compliment several times before, we at first concluded to be nothing but a morning salute.[39]

Colonel Woodford had a similar reaction.

After reveille beating, two or three great guns, and some musquetry were discharged from the enemy's fort, which, as it was not an unusual thing, was but little regarded.[40]

Behind the rebel breastworks, however, stood sixty odd Virginians with orders to hold their fire until the enemy approached within fifty yards of the works.[41] To their front

---

[38] Ibid.
[39] Peter Force, ed., "Major Spotswood to a Friend in Williamsburgh, December 9, 1775," *American Archives, Fourth Series*, Vol. 4, 224.
[40] Clark, ed., "Colonel Woodford to Edmund Pendleton, December 10, 1775," *Naval Documents of the American Revolution*, Vol. 3, 39-40.
[41] Force, ed., "Major Spotswood to a Friend in Williamsburgh, December 9, 1775," *American Archives, Fourth Series*, Vol. 4, 224.

were more than five times their number of enemy troops with two cannon. The British cannon and about 200 Tory and black troops halted in front of the narrow portion of the causeway, 200 yards from the rebel works, and engaged the rebels from a distance.[42] Captain Samuel Leslie remained with this force while Captain Charles Fordyce led approximately 120 British regulars forward, six abreast, onto the narrow portion of the causeway and straight at the rebel works.[43]

When they approached to within fifty yards of the rebel breastworks they were met with a ferocious volley that, according to one witness, *"threw them into some confusion."*[44] Captain Fordyce rallied the redcoats forward but was soon riddled by buck and ball and fell just short of the rebel works. A British midshipman witnessed the attack and recalled that the rebel fire, *"was so heavy, that, had we not retreated as we did, we should every one have been cut off."*[45] He described the attack in detail.

> *Figure to yourself a strong breast-work built across a causeway, on which six men only could advance a-breast; a large swamp almost surrounding them, at the back of which were two small breast-works to flank us in our attack on their entrenchments. Under these disadvantages it was impossible to succeed....*[46]

---

[42] Clark, ed., "Letter to Pinkney, December 20, 1775," *Naval Documents of the American Revolution*, Vol. 3, 186-189.
[43] Clark, ed., "Letter from a Midshipman on *H.M. Sloop Otter*, December 9, 1775," *Naval Documents of the American Revolution*, Vol.3, 29.
[44] Clark, ed., "Letter to Pinkney, December 20, 1775," *Naval Documents of the American Revolution*, Vol. 3, 186-189.
[45] Clark, ed., "Letter from a Midshipman on *H.M. Sloop Otter*, December 9, 1775," *Naval Documents of the American Revolution*, Vol. 3, 29.
[46] Ibid.

Captain Robert Kidder Meade of the 2nd Virginia described the effect of their fire upon the attacking enemy.

> *The scene, when the dead and wounded were bro't off, was too much; I then saw the horrors of war in perfection, worse than can be imagin'd; 10 and 12 bullets thro' many; limbs broke in 2 or 3 places; brains turning out. Good God, what a sight!*[47]

Captain Fordyce and twelve British privates lay dead in front of the rebel breastworks and nearly a score of wounded redcoats were taken prisoner.[48] The remainder, some of who were also wounded, withdrew to the cannon and made a brief stand, but retreated back to the fort under heavy fire with the rest of the troops.

Captain Leslie, distraught over the loss of his men as well as a nephew, Lieutenant Peter Leslie, abandoned the fort at nightfall and marched back to Norfolk, where he reportedly declared that, "*no more of his troops should be sacrificed to whims.*"[49]

A jubilant Colonel Woodford described the engagement to the Committee of Safety as, "*a second Bunker's Hill affair, in miniature; with this difference, that we kept our post, and had*

---

[47] Charles Campbell, ed., "Richard Kidder Meade to Theodorick Bland Jr., December 18, 1775," *The Bland Papers*, Vol. 1, (1840), 38-39.

[48] Clark, ed., "Colonel Woodford to Edmund Pendleton, December 10, 1775, "*Naval Documents of the American Revolution*, Vol. 3, 39-40.

[49] Clark, ed., "Letter from the Virginia Committee of Safety, December 16, 1775, "*Naval Documents of the American Revolution*, Vol. 3, 132.

only one man wounded in the hand."[50] Upon his inspection of the abandoned fort Woodford added that,

> *From the vast effusion of blood on the bridge, and in the fort, from the accounts of the sentries, who saw many bodies carried out of the fort to be interred and other circumstances, I conceive their loss to be much greater than I thought it yesterday, and the victory to be complete.*[51]

The victory was indeed complete. The 14th regiment had lost approximately half its men in killed and wounded and the fear that spread among Dunmore's supporters in Norfolk forced him to abandon the town for the safety of ships in the harbor.[52]

Continental troops from North Carolina arrived a few days after the battle and their commander, Colonel Robert Howe, assumed command by virtue of his continental rank. Colonel Woodford was likely displeased by this situation, but found some solace in the Continental Congress's decision on December 28th, to place six Virginia regiments, including his 2nd Virginia as well as the 1st Virginia Regiment, onto the continental establishment.[53]

---

[50] Clark, ed., "Colonel Woodford to Edmund Pendleton, December 10, 1775, *Naval Documents of the American Revolution*, Vol. 3, 39-40.
[51] Ibid.
[52] Clark, ed., "Lord Dunmore to Lord Dartmouth, December 13, 1775," *Naval Documents of the American Revolution*, Vol. 3, 141.
[53] Ford, ed., "Proceedings of Congress, December 28, 1775," *Journals of the Continental Congress*, Vol. 3, 463.

Battle of Great Bridge

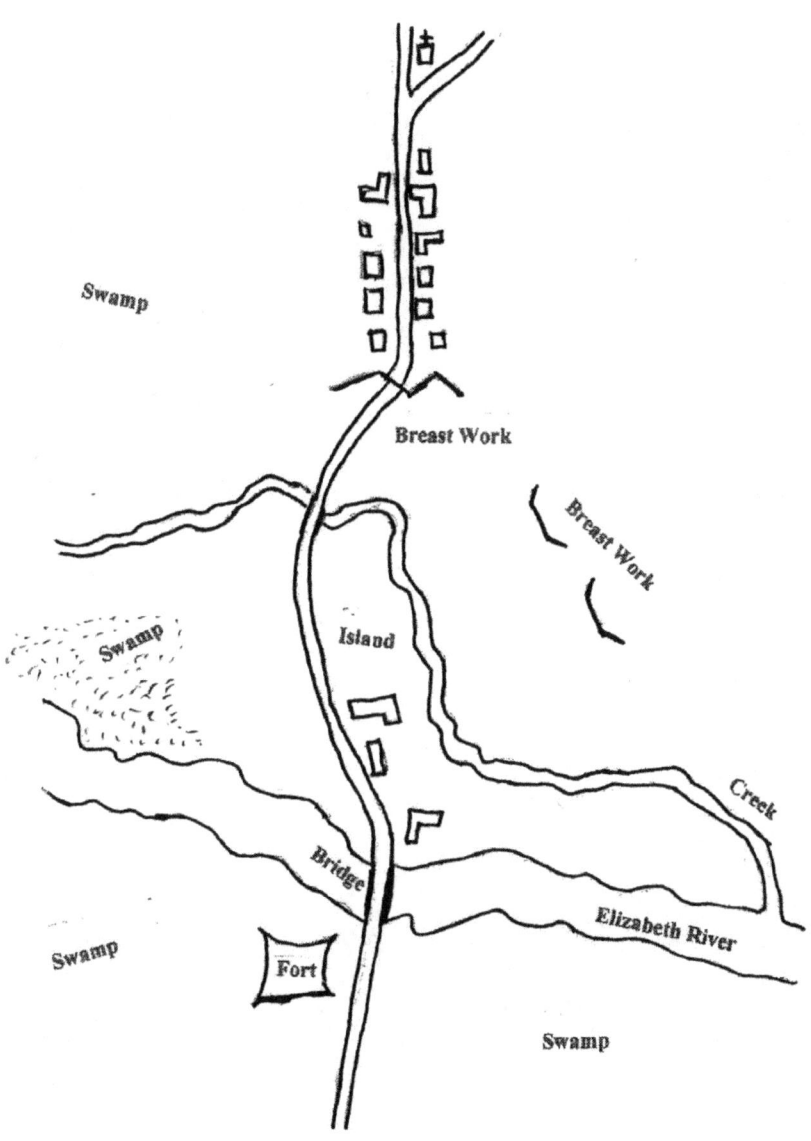

Chapter Three

The Virginia Continental Line Forms

December 1775 – March 1776

The 4th Virginia Convention, meeting in Williamsburg in December 1775, also recognized the need to expand Virginia's military and began debate on a resolution to raise six additional regiments of regulars to support the 1st and 2nd Regiments two weeks before Congress authorized the six continental regiments for Virginia.[1] The new regiments were to consist of ten companies of sixty-eight rank and file each who would serve two year enlistments. The 1st and 2nd Regiments were also to be increased to ten companies and the troops of the sixteen original companies of regulars raised in the fall of 1775 were encouraged to reenlist in their two regiments under the new terms of the six additional regiments.[2]

Debate over the specific details continued into January and on January 11, 1776 the 4th Virginia Convention authorized six new regiments, plus one additional regiment made up of seven companies to defend the Eastern Shore of Virginia.[3] This brought the number of Virginia's regiments of regular troops to

---

[1] Robert L. Scribner and Brent Tarter, eds., "Proceedings of the 4th Virginia Convention, December 13, 1775," *Revolutionary Virginia: The Road to Independence,* Vol. 5, (University Press of Virginia, 1979), 128.
[2] Ibid. and Hening, ed., *Statutes at Large*, Vol. 9, 91.
[3] Scribner and Tarter, eds., "Proceedings of the 4th Virginia Convention, January 11, 1776," *Revolutionary Virginia: The Road to Independence,* Vol. 5, 383, and William W. Hening, ed., *Statutes at Large*, Vol. 9, 76-77.

nine, three more than the Continental Congress was willing to pay for.

All but the last two regiments, the 8th and 9th Regiments, were to consist of seven line (musket) companies and three rifle companies, whom the delegates noted were, *"to act as light infantry."*[4] The 9th Virginia Regiment on the Eastern Shore was allotted two rifle companies for that purpose and five line companies while all of the troops in the 8th Regiment, recruited from the Shenandoah Valley and further west, carried rifles.[5]

The addition of approximately 4,500 regular troops and officers in Virginia required a substantially greater recruitment effort, so the Convention instructed every county (instead of the sixteen districts formed in 1775) to raise at least one full company of men and for the larger counties, two companies. The appointment of company level officers was left to the individual counties; the convention appointed the regimental staff officers such as colonels, lieutenant colonels, and majors.[6]

To encourage the enlistment of the large number of German settlers in the Shenandoah Valley, the 8th Virginia Regiment was designated the German Regiment. This turned out to be somewhat misleading, however, for although the regiment did eventually consist of a large number of German speaking troops, Irish and Scottish troops were also well represented in the unit.[7]

---

[4] Hening, ed., *Statutes at Large*, Vol. 9, 76-77.
[5] Ibid, and Frank E. Grizzard Jr., ed., "General Peter Muhlenberg to General Washington, February 23, 1777," *The Papers of George Washington, Revolutionary War Series*, Vol. 8, (Charlottesville: University Press of Virginia, 1998), 428.
[6] Hening, ed., *Statutes at Large*, Vol. 9, 77.
[7] Brent Tarter, ed., "General Charles Lee to Edmund Pendleton, May 24, 1776," *Revolutionary Virginia" The Road to Independence*, Vol. 7 Part One, (University Press of Virginia, 1983), 248-249.

Disappointed that the Continental Congress had agreed to take only six of its provincial regiments onto the continental establishment, the 4th Convention appealed to and cajoled Congress to take all nine.

> *The Convention had hoped that Congress would have supposed them competent Judges of the number of Forces necessary to the security of this Colony* [which the Convention determined was nine provincial regiments] *and considering the large Proportion of the Continental Expences which* [Virginia's] *Inhabitants are to pay that* [Congress] *would without Hesitation have taken all the regular forces found necessary for their Defence into continental Pay.*[8]

The Convention also desired, out of fairness to the officers of the 1st and 2nd Regiments, that those two regiments stand first in the arrangement of the Virginia Continental Line and that the continental commissions of the officers of these two regiments be back dated to November 1st, 1775 in recognition of the, "*laborious and painful service*," the officers had endured the previous fall.[9] This point was of particular interest to the officers in question for it impacted their standing within the officer's corps, making them senior to all of the new officers of equal rank in the additional Virginia regiments.

While Virginia's leaders waited for Congress to consider the 4th Convention's appeal to take all nine provincial regiments

---

[8] Scribner and Tarter, eds., "Proceedings of the 4th Virginia Convention, January 10, 1776," *Revolutionary Virginia: The Road to Independence,* Vol. 5, 372-373.
[9] Ibid., 373.

into the continental army, county leaders throughout the Old Dominion undertook the challenge of raising troops to complete their allotted companies. Hundreds of miles to the north, Virginia's first continental troops, the riflemen of Captain Daniel Morgan's and Captain Hugh Stephenson's companies, struggled through a difficult winter in Quebec and Boston.

Virginia's Continental Rifle Companies

Captain Daniel Morgan and his Virginia riflemen spent November and December outside the walled city of Quebec with Colonel Benedict Arnold's depleted force waiting for reinforcements. When they finally arrived in December under General Richard Montgomery, they failed to even replace the number of men Arnold lost on his march to Quebec.

General Montgomery assumed command of the American troops at Quebec and attempted to take the fortified town by storm on December 31st. Advancing in the van of the attack under cover of darkness in a blizzard, Captain Morgan and his Virginia riflemen fought their way past one enemy strongpoint but became trapped in the lower town of Quebec. Forced to surrender, Captain Morgan and many of his men spent the next seven months as prisoners of war until they were paroled and allowed to return to Virginia in August.

Outside of Boston, General Washington's ill supplied army suffered through the winter and the stalemate with the British in Boston. Captain Hugh Stephenson and his Virginia riflemen were posted in the Roxbury lines facing Boston neck. Stephenson had been placed in command of the two Maryland rifle companies as well as his own and together the three companies acted as one unit.[10] Like the rest of Washington's army, Captain Stephenson and his riflemen saw little action over the long winter and waited eagerly for the advent of spring.

The Virginia Continental Line Takes Form

The situation in southeastern Virginia at the end of 1775 was also stalemated. Colonel Woodford's force of 2nd Virginia troops and Culpeper minutemen, reinforced by several companies from the 1st Virginia as well as a large detachment of North Carolinians under Colonel Robert Howe (who by virtue of his continental commission assumed command of the entire force) occupied the abandoned city of Norfolk. Just offshore in the Elizabeth River sat Governor Dunmore with his supporters and the survivors of the 14th Regiment aboard an assortment of vessels protected by several British warships.

On New Year's Day, after several days of provocation by the Virginia and Carolina rebels, British warships bombarded Norfolk and set fire to buildings along the shore used as shelter by rebel sentries. The fire quickly spread and, aided by the rebels themselves who took the opportunity to torch many more buildings in what they viewed was a Tory infested city, much of Norfolk was burned to the ground. The ruins made for uncomfortable accommodations for the Virginia and North Carolina troops so Colonel Howe and Colonel Woodford withdrew the bulk of their troops to Suffolk at the end of

---

[10] Philander D. Chase, ed., "General Washington to Samuel Washington, "September 30, 1775," *The Papers of George Washington, Revolutionary War Series,* Vol. 2, (Charlottesville: University Press of Virginia, 1987), 73, and "General Orders, March 13, 1776," *The Papers of George Washington, Revolutionary War Series,* Vol. 3, (Charlottesville: University Press of Virginia, 1988), 458.

January. Dunmore and his supporters remained aboard their floating town relatively undisturbed, inactive, and hopeful that British help would soon arrive.

Efforts by Virginia's delegates in the Continental Congress to convince that body to increase Virginia's six continental regiments to nine initially met with failure.[11] Virginia's leaders were concerned that by not taking all of the colony's regiments onto continental service, Congress would bypass the 1st and 2nd Regiments, leaving them as provincial units and their officers inferior in rank to the continental officers of the six new regiments being raised in Virginia.

To avoid such a disservice to the officers of the 1st and 2nd Regiments, the Virginia Committee of Safety sent a list of officers it wished to serve in the six authorized continental regiments, which the committee identified as the 1st through 6th Regiments, and on February 13th, Congress approved the list.[12] Continental commissions for the appointed officers were then sent from Philadelphia to Williamsburg to be signed and delivered to the regimental officers listed by the Committee of Safety.

Colonel Patrick Henry Resigns

On February 28th, the Committee summoned Patrick Henry to receive his newly arrived continental commission as colonel of the 1st Virginia Regiment, but upon reading it he

---

[11] Ford, ed., "Proceeding of the Continental Congress, February 13, 1776," *Journals of the Continental Congress*, Vol. 4, 132.
[12] Ibid., and Purdie, "March 15, 1776," *Virginia Gazette Supplement*, 2.

declined to accept it and offered his resignation from the army instead.[13]

Henry informed his brother-in-law, Samuel Meredith that his resignation was the result of the Committee of Safety's preference for subordinate officers over himself.[14] This was likely a reference to the committee's earlier decision to send Colonel Woodford and the 2nd Virginia Regiment to Great Bridge instead of Henry and the 1st Virginia. Numerous slights by Colonel Woodford as well as the Committee of Safety likely also contributed to Henry's decision.

Edmund Pendleton, the President of the Committee of Safety and a longtime rival of Henry, speculated that Henry's resignation was more the result of wounded pride and the loss of, "*the Fringe of Commander in chief,*" which slipped from Henry when Congress appointed him as colonel of the 1st Virginia instead of brigadier general of all of Virginia's continental forces. As the ranking military commander in Virginia, Henry had expected that honor, but his commission revealed that he had not been promoted to brigadier general. Congress instead selected Andrew Lewis, a veteran of both the French and Indian War and Lord Dunmore's expedition against the Shawnee Indians in 1774, to serve as Virginia's brigadier-general.[15]

Whatever the reason for his resignation, Patrick Henry's departure from the army freed him to resume his leadership in civil affairs, something he excelled at. Henry's political

---

[13] Purdie, "March 15, 1776," *Virginia Gazette Supplement*, 2.
[14] Robert L. Scribner and Brent Tarter, eds., *Revolutionary Virginia: The Road to Independence*, Vol. 6, (University Press of Virginia, 1981), 5.
[15] Ibid., and Ford, ed., "Proceeding of the Continental Congress, March 1, 1776," *Journals of the Continental Congress*, Vol. 4, 181.

leadership led to his election as Virginia's first post-colonial governor just four months after his resignation from the army.

The news of Henry's resignation from the army prompted a near mutiny in the ranks of the 1st Virginia Regiment; the troops went into, *"deep mourning"* and gathered, under arms at Henry's lodging in Williamsburg to address him.[16] Henry graciously thanked those assembled for their support and then attended a farewell dinner with his former officers at the Raleigh Tavern in his honor. He was forced to postpone his departure from the capital, however, when word spread of, *"some uneasiness getting among the soldiery, who assembled in a tumultuous manner, and demanded their discharge, declaring their unwillingness to serve under any other commander."*[17] Henry spent most of the evening with the troops, *"visiting the several barracks, and* [using] *every argument in his power with the soldiery to lay aside their imprudent resolution, and continue in the service...."*[18] His efforts eventually succeeded and the disgruntled soldiers reluctantly accepted Henry's departure.

Recruitment of Continental Regiments

County committees throughout Virginia appointed the captains, lieutenants, and ensigns in January and February who were to command and care for the young men raised in their counties. As leaders within their respective counties, each appointed officer was expected to recruit a specific number of men to "earn" his commission. Captains were to recruit 28

---

[16] Purdie, "March 1, 1776," *Virginia Gazette*, 3.
[17] Ibid.
[18] Ibid.

men, first lieutenants 21 men, second lieutenants, 16 men, and ensigns 9 men. This came to a total of 74 soldiers per company, consisting of 68 rank of file, 4 sergeants, and 2 musicians (a drummer and fifer).[19]

Recruits received a 20 shilling bounty for their two year enlistment and had to be at least *"five foot four inches high, healthy, strong made & well limbed, not deaf or Subject to fits."*[20] No servant whatsoever was allowed to serve unless he was apprenticed and his master provided written permission for him to serve.[21]

The Virginia Committee of Safety, led by Edmund Pendleton, acted as Virginia's civil government during the winter and spring of 1776 and coordinated the recruitment effort and the disposition of troops. In the fall of 1775, the 1st and 2nd Virginia Regiments had been instructed to muster in Williamsburg. With four times the number of troops to raise in 1776, Virginia's leaders worried about the ability of the capital to accommodate so many troops. They were also concerned that Virginia's many navigable rivers and creeks made its several peninsulas vulnerable to attack. The Committee of Safety opted therefore to disperse the new regiments between the four main rivers in the eastern part of Virginia, the James, York, Rappahannock, and Potomac Rivers.[22]

---

[19] Hening, ed., *Statutes at Large*, Vol. 9, 77-78.
[20] Scribner and Tarter, ed., "Orders for Recruiting Regular Soldiers, January 22, 1776" *Revolutionary Virginia: Road to Independence*, Vol. 6, 15-16.
[21] Ibid.
[22] Scribner and Tarter, ed., "Proceedings of the Virginia Committee of Safety, February 10, 1776" *Revolutionary Virginia: Road to Independence*, Vol. 6, 85.

Although the new regiments had not yet been assigned their companies, the Committee of Safety decided on February 10th, to post the 1st and 6th Regiments between the James and York River (at Williamsburg and Hampton), the 2nd and 7th Regiments between the York and Rappahannock Rivers (at Gloucester Courthouse), the 3rd and 5th Regiments between the Rappahannock and Potomac Rivers (at Dumfries and Richmond Courthouse) and the 4th and 8th Regiments on the south side of the James River (at and around Suffolk). The 9th Regiment, with only seven companies, was to remain on the Eastern shore.[23]

Virginia's thirteen largest counties were instructed to raise two new companies of regulars. Thirty-three other counties raised one new company each and the smallest remaining counties, as well as the cities of Williamsburg and Norfolk, were grouped together to raise 13 additional companies. All of these companies except five were allotted to the seven new regiments being raised, the 3rd through 9th Regiments. Five of the new companies, however, were attached to the 1st and 2nd Regiments to bring them up to ten companies each.[24]

Along with the bounty provided to all enlistees, Virginia's leaders agreed to provide the recruits, at the colony's expense, with a hunting shirt, pair of leggings, and binding for a hat. They were also to be furnished with one good musket and bayonet, a cartouch box or pouch, and a canteen. As the colony did not possess nearly enough muskets to arm all of the new recruits, soldiers were encouraged to bring, *"the best gun of any sort they can procure"* for which they received 20 shillings a

---

[23] Ibid.
[24] Hening, ed., *Statutes at Large*, Vol. 9, 77-78.

year as an inducement.[25] Blankets were also in short supply so the men were encouraged to bring their own.[26] Soldiers in need of additional clothing (breeches, shirts, shoes, hats) were provided with them, but the cost was deducted from their pay.[27]

The pay for the officers and men was modest. In an age when 20 to 30 pounds was a typical annual household income for many Virginians, Virginia's regulars were reasonably paid. Colonels received 17.5 shillings per day which came to over 26 pounds a month and nearly 315 pounds a year. Captains received 6 shillings a day which amounted to 9 pounds per month and 108 pounds for a year. Lieutenants received 2/3 of a captain's pay (4 shillings a day), sergeants received half of a lieutenant's pay (2 shillings a day) and private soldiers received 1 shilling 4 pence a day, which amounted to 2.5 pounds a month or 30 pounds a year.[28]

This pay scale lasted but a few months, replaced by a new continental pay scale when the Virginia regulars were taken into continental service. Under the continental scale, colonels received $50 a month or $600 a year, captains received $26.67 a month or $320 a year, lieutenants $18 a month or $216 a year, sergeants, $8 a month or $96 a year and privates $6.67 a month or $80 a year.[29] Although the continental pay appeared to be much more generous, it is difficult to assess its true value as it was based on continental dollars, not English pounds, and the

---

[25] Ibid., 81-82.
[26] Ibid.
[27] Ibid., 82.
[28] Ibid., 82-83.
[29] Ford, ed., "Proceedings of the Continental Congress, June 29, 1775 and November 4, 1776," *Journals of the Continental Congress*, Vol. 2, 220 and Vol. 3, 322.

value of those dollars was anyone's guess, especially as the war progressed.

The Committee of Safety exercised its authority over the winter to assign Virginia's seventy-two new companies of regulars, comprised of troops enlisted for two years, to their respective regiments. Instructions were sent to each county in February and March detailing which regiment their company or companies were assigned to and where and when they were to rendezvous with the rest of their regiment. The Committee of Safety also sent blank continental commissions to the counties to be filled in and delivered by county leaders to the company officers once their companies were reviewed and sworn into service.[30]

Although most of the company officers received their commissions from their county committees, the regimental officers (colonels, lieutenant colonels, and majors) went to the capital to receive their continental commissions directly from the Committee of Safety who in turn, received them from the Continental Congress. Most the officers stayed briefly in Williamsburg to subscribe to the Articles of War (rules and regulations for the army) and then continued on with their commissions in hand to join their troops at their assigned stations in Prince William, Richmond, and Gloucester Counties as well as in Suffolk.

The 1st Regiment and the 6th Regiment were posted in Williamsburg and although several of the 1st Regiment's companies had been detached to Colonel Woodford's force in Suffolk and several more were posted in Hampton, the

---

[30] Scribner and Tarter, eds., "Proceedings of the Committee of Safety, March 13, 1776," *Revolutionary Virginia, The Road to Independence*, Vol. 6, 204.

remainder of the regiment spent the winter in Williamsburg. They were still there when the troops of the 6th Virginia began to arrive in March.

There were actually three regiments posted in Suffolk during the spring of 1776. After their victory at Great Bridge and the destruction of Norfolk, the 2nd Virginia Regiment spent the winter in Suffolk and in the spring was joined by companies from the 4th and 8th Regiments. This left the 7th Virginia to guard the area between the York and Rappahannock Rivers alone. They mustered at Gloucester Courthouse in the spring.

The 5th Virginia Regiment mustered at Richmond Courthouse (today Warsaw, Virginia) on the north side of the Rappahannock River and the 3rd Virginia mustered in Dumfries and Alexandria on the Potomac River. The 9th Virginia remained on the Eastern Shore to protect that region.[31]

First Virginia Regiment of 1776

The resignation of Colonel Patrick Henry from the 1st Regiment at the end of February led to the elevation and promotion of Colonel William Christian, who assumed command of the regiment in March. Major Francis Epps was promoted to Lieutenant Colonel in mid-March and Captain John Green elevated to Major.[32]

The eight original companies of the regiment that were raised in 1775 still had six months of service on their original enlistments. They included Captain John Green's company from the Culpeper district, Captain John Markham's company

---

[31] Hening, ed., *Statutes at Large*, Vol. 9, 77-78.
[32] Scribner and Tarter, ed., "Proceedings of the Continental Congress, March 18, 1776," *Revolutionary Virginia: Road to Independence*, Vol. 6, 224.

from the Amelia district, Captain John Sayres's company from the Gloucester district, Captain William Davies's company from the Princess Anne district, Captain Robert Ballard's company from the Mechlenburg district, Captain John Fleming's company from the Henrico district, Captain William Campbell's company from the Pittsylvania district, and Captain George Gibson's company from the West August district.[33] The two additional companies attached to the 1st Regiment in 1776 were raised in York County (under Captain Thomas Nelson Jr.) and Williamsburg (under Captain Edmund Dickenson).[34] These troops enlisted for two years. Several companies of the 1st Regiment had seen action along the James River and at Norfolk, and all except the last two had six months of experience in the army.

Second Virginia Regiment of 1776

Colonel William Woodford remained the commander of the 2nd Virginia Regiment. When he returned to his home in Caroline County in March to restore his health, command fell to Lieutenant Colonel Charles Scott. He was assisted by Major Alexander Spotswood.

---

[33] Tarter, ed., "Orders for October 21, 1775" "The Orderly Book of the Second Virginia Regiment," *The Virginia Magazine of History and Biography*, Vol. 85, No. 2, 170-171, and Sanchez-Saavedra, ed., *A Guide to Virginia Military Organizations in the American Revolution, 1774-1787*, 29-31.

Note: The companies were raised among districts drawn up by the 3rd Virginia Convention that were comprised of several counties. The respective captains above were not necessarily from the county that follows their names, but rather, from the district that follows their name.

[34] Sanchez-Saavedra, ed., *A Guide to Virginia Military Organizations in the American Revolution, 1774-1787*, 31.

Like most of the companies of the 1st Virginia, the seven original companies of the 2nd Virginia that were raised in 1775 had six more months of service left. They included Captain Morgan Alexander's company from the Berkeley district, Captain William Fontaine's company from the Buckingham district, Captain William Taliaferro's company from the Caroline district, Captain George Johnston's company from the Prince William district, Captain George Nicholas's company from the Elizabeth City district, Captain Richard Parker's company from the Lancaster district, and Captain Richard Kidder Meade's company from the Southampton district.[35]

The three additional companies attached to the 2nd Virginia Regiment in 1776 were raised in Prince George County under Captain Buller Claiborne, Caroline County under Captain Samuel Hawes and Amelia County under Captain Wood Jones.[36]

The soldiers of the 2nd Regiment had seen the bulk of the fighting among Virginia's regulars and although the Committee of Safety wished to post the regiment between the Rappahannock and York Rivers, the 2nd Virginia remained in Suffolk for most of the spring, waiting for the companies of the 4th and 8th Regiments to arrive in sufficient number to allow them to take their new post above the York River.

---

[35] Tarter, ed., "Orders for October 21, 1775" "The Orderly Book of the Second Virginia Regiment," *The Virginia Magazine of History and Biography*, Vol. 85, No. 2, 171 and Sanchez-Saavedra, ed., *A Guide to Virginia Military Organizations in the American Revolution, 1774-1787*, 36.

[36] Sanchez-Saavedra, ed., *A Guide to Virginia Military Organizations in the American Revolution, 1774-1787*, 36.

Third Virginia Regiment of 1776

Colonel Hugh Mercer of Fredericksburg commanded the 3rd Virginia Regiment.[37] His experience in the French and Indian War in command of Pennsylvanian troops nearly resulted in his appointment to command the original 1st Virginia, but prejudice by some in the 3rd Virginia Convention towards Mercer's Scottish background and the popularity of Patrick Henry resulted in Mercer being passed over. He was joined by Lieutenant Colonel George Weedon, a successful tavern owner in Fredericksburg (and Mercer's brother-in-law through marriage) and Major Thomas Marshall of Fauquier County, the father of future Supreme Court Chief Justice John Marshall.[38]

The ten companies that comprised the 3rd Virginia Regiment in 1776 were raised in the counties of Prince William (Captain Andrew Leitch and Captain Philip Richard Francis Lee), Fairfax (Captain John Fitzgerald), Loudoun (Captain Charles West), Culpeper (Captain John Thornton), King George (Captain Gustavus Brown Wallace), Stafford (Captain William Washington), Spotsylvania (Captain William McWilliams), Fauquier (Captain John Chilton) and Louisa (Captain Thomas Johnson Jr.).[39]

---

[37] Scribner and Tarter, eds., "Proceedings of the 4th Virginia Convention, January 11, 1776," *Revolutionary Virginia: The Road to Independence,* Vol. 5, 383.

[38] Scribner and Tarter, eds., "Proceedings of the 4th Virginia Convention, January 12, 1776," *Revolutionary Virginia: The Road to Independence,* Vol. 5, 391-392.

[39] Sanchez-Saavedra, ed., *A Guide to Virginia Military Organizations in the American Revolution, 1774-1787,* 39.

Fourth Virginia Regiment of 1776

Colonel Adam Stephen from Berkeley County, who served with Colonel Washington as a lieutenant in the French and Indian War, commanded the 4th Virginia Regiment. He was joined by Lieutenant Colonel Isaac Read of Charlotte County and Major Robert Lawson of Henrico County.[40]

The ten companies that comprised the 4th Virginia Regiment in 1776 were raised in the counties of Berkeley (Captain Isaac Beall), Prince Edward (Captain John Morton), Charlotte (Captain John Brent), Southampton (Captain Thomas Ridley), Sussex (Captain Nathaniel Mason), Brunswick (Captain James Lucas). Nansemond (Captain John Washington), and three companies from Isle of Wight and Surry Counties combined under Captain John Watkins Jr., Captain Thomas Mathews and Captain Arthur Smith.[41]

Fifth Virginia Regiment of 1776

Colonel William Peachy of Richmond County briefly commanded the 5th Virginia Regiment at its formation in 1776. He was joined by Lieutenant Colonel William Crawford of Berkeley County and Major Josiah Parker of Isle of Wight County.[42]

---

[40] Scribner and Tarter, eds., "Proceedings of the 4th Virginia Convention, January 12, 1776," *Revolutionary Virginia: The Road to Independence*, Vol. 5, 390-392.

[41] Sanchez-Saavedra, ed., *A Guide to Virginia Military Organizations in the American Revolution, 1774-1787*, 43.

[42] Scribner and Tarter, eds., "Proceedings of the 4th Virginia Convention, January 12, 1776," *Revolutionary Virginia: The Road to Independence*, Vol. 5, 390-392.

The ten companies that comprised the 5th Virginia Regiment in 1776 were raised in the counties of Lancaster, (Captain Burgess Ball), Orange (Captain George Stubblefield), Henrico (Captain John Pleasants), Northumberland (Captain Thomas Gaskins), Bedford (Captain Gross Scruggs and Captain Henry Terrill), Chesterfield (Captain Ralph Faulkner), Hanover (Captain Richard Clough Anderson), Loudoun (Captain Andrew Russell), and Richmond (Captain Henry Fauntleroy).[43]

Sixth Virginia Regiment of 1776

Colonel Mordecai Buckner of Caroline County commanded the 6th Virginia Regiment. He was joined by Lieutenant Colonel Thomas Elliot of King William County and Major James Hendricks of Fairfax County. [44]

The ten companies that comprised the 6th Virginia Regiment in 1776 were raised in the counties of Lunenburg (Captain James Johnson), Spotsylvania (Captain Oliver Towles), Buckingham (Captain Thomas Patterson), Mechlenburg (Captain Samuel Hopkins), Charles City (Captain William Gregory), Prince George (Captain Thomas Ruffin), Amherst (Captain Samuel Jordan Cabell), New Kent (Captain Thomas Massie), Pittsylvania (Captain Thomas Hutchings), and Dinwiddie, (Captain John Jones).[45]

---

[43] Sanchez-Saavedra, ed., *A Guide to Virginia Military Organizations in the American Revolution, 1774-1787*, 45-46.
[44] Scribner and Tarter, eds., "Proceedings of the 4th Virginia Convention, January 12, 1776," *Revolutionary Virginia: The Road to Independence*, Vol. 5, 390-393.
[45] Sanchez-Saavedra, ed., *A Guide to Virginia Military Organizations in the American Revolution, 1774-1787*, 49.

Seventh Virginia Regiment of 1776

Colonel William Daingerfield of Spotsylvania County commanded the 7th Regiment. He was joined by Lieutenant Colonel Alexander McClanahan of Augusta County and Major William Nelson of King William County.[46]

The ten companies that comprised the 7th Virginia Regiment in 1776 were raised in the counties of King and Queen (Captain Gregory Smith), King William (Captain Holt Richeson), Cumberland (Captain Charles Fleming), Essex and Middlesex (Captain John Webb), Halifax (Captain Nathaniel Cocke), Gloucester (Captain Charles Tomkies), Albemarle (Captain Matthew Jouett), Botetourt (Captain Thomas Posey), Fincastle (Captain Joseph Crockett) and Orange, (Captain Joseph Spencer).[47]

Eighth Virginia Regiment of 1776

The 8th Virginia Regiment, designated as the German Regiment to encourage the many German speaking settlers of the Shenandoah Valley to enlist in it, was commanded by Colonel Peter Muhlenberg.[48] Born and raised in Pennsylvania, Muhlenberg had only been in Virginia for four years, serving as a minister to a parish in the Shenandoah Valley. His brief stint

---

[46] Scribner and Tarter, eds., "Proceedings of the 4th Virginia Convention, January 12, 1776," *Revolutionary Virginia: The Road to Independence,* Vol. 5, 391-393.

[47] Sanchez-Saavedra, ed., *A Guide to Virginia Military Organizations in the American Revolution, 1774-1787,* 52.

[48] Scribner and Tarter, eds., "Proceedings of the 4th Virginia Convention, January 12, 1776," *Revolutionary Virginia: The Road to Independence,* Vol. 5, 391.

in the 60th British Regiment in 1767 as a secretary, combined with the esteem that the inhabitants of Dunmore County held for the twenty-nine year old, was apparently enough to warrant his appointment to command the German Regiment. He was joined by Lieutenant Colonel Abraham Bowman of Dunmore County and Major Peter Helphinstine of Frederick County, two additional German speaking officers.[49]

The ten companies that comprised the 8th Virginia Regiment in 1776 were all rifle companies raised in the counties of Augusta (Captain John Stevenson and Captain David Stephenson), Dunmore (Captain Jonathan Clark and Captain Richard Campbell), Culpeper (Captain George Slaughter), Berkeley (Captain William Darke), Hampshire (Captain Abel Westfall), Frederick (Captain Thomas Berry), Fincastle (Captain James Knox), and West Augusta (Captain William Croghan.[50]

Ninth Virginia Regiment of 1776

Colonel Thomas Fleming of Chesterfield County commanded the 9th Virginia Regiment, posted on the Eastern Shore. He was joined by Lieutenant Colonel George Mathews of Augusta County and Major Matthew Donavan of Northampton County.[51]

The seven companies that comprised the 9th Virginia Regiment in 1776 were raised primarily on the Eastern Shore

---

[49] Ibid., 391.
[50] Sanchez-Saavedra, ed., *A Guide to Virginia Military Organizations in the American Revolution, 1774-1787*, 55.
[51] Scribner and Tarter, eds., "Proceedings of the 4th Virginia Convention, January 12, 1776," *Revolutionary Virginia: The Road to Independence*, Vol. 5, 391-393.

in Accomack County. Captains John Cropper, Levin Joynes, Thomas Davis, and Thomas Snead, commanded companies from that county. A company of riflemen from Albemarle County under Captain Thomas Walker, a company from Goochland County under Captain Samuel Woodson and another company from Gloucester County under Captain John Hayes completed the seven companies of the 9th Regiment in the spring of 1776. Three additional companies from Accomack County were added to the regiment in the summer under the command of Captains George Gilchrist, Thomas Parramore, and John Poulson, bringing the regiment to ten companies like the rest of the Virginia Continental Line.[52]

Patrick Henry's refusal to accept his commission and subsequent departure from the army left a pall of resentment that extended into spring among his former troops.[53] It was thus a surly group of disgruntled soldiers encamped in Williamsburg who greeted Colonel Hugh Mercer when he arrived in March to receive his continental commission as colonel of the 3rd Regiment. With Colonel Woodford away on leave, Colonel Mercer found himself the ranking officer in town and temporarily assumed command of all the troops in the capital.

Colonel Mercer immediately recognized the need for greater discipline, both among the rank and file and the officers, and issued orders accordingly.

All Orders with Respect to the [troops] *are to be Frequently Read and Explained by the Captains of Companies to their men on the Parade, so that Every Man*

---

[52] Sanchez-Saavedra, ed., *A Guide to Virginia Military Organizations in the American Revolution, 1774-1787*, 59.
[53] Purdie, "March 1, 1776," *Virginia Gazette*, 3.

may Fully understand them. The officers will always Treat the Soldiers With the utmost Humanity and Tenderness, and the Soldiers and Non-Comm's'd officers to pay Due Respect to their officers. A Soldier, who Knows and inclines to do his Duty, will never pass Without moving his Hatt. Every morning, after Divine Service, the Companys are to be Drill'd, and is to be Expected that the officers of Each Comp'y will exert their utmost abilities to form their Men in Due Military order.[54]

Edmund Pendleton acknowledged Mercer's progress with the troops but also noted some resistance to his efforts:

Colonel Mercer has done great things towards a Reform which has given great pleasure to the Judicious, but I understand has produced a Court of Enquiry into his Conduct....[55]

The specific incident that sparked an inquiry into Mercer's conduct involved Captain George Gibson's company of Augusta County riflemen of the 1st Virginia (who were under Lieutenant William Lynn's command in Gibson's absence). These frontiersmen bristled at Mercer's efforts to reign in their behavior, and when two of them were lightly punished by a court martial for behavior that Colonel Mercer believed was

---

[54] R. A. Brock, ed., "Orderly Book of Capt. George Stubblefield, March 11, 1776," *Miscellaneous Papers…in the Collections of the Virginia Historical Society*, (Richmond, VA: Virginia Historical Society, 1887), 150.

[55] David John Mays, ed., "To William Woodford, March 16, 1776," *The Letters and Papers of Edmund Pendleton, 1734-1803*, Vol. 1, (Charlottesville, VA: University Press of Virginia, 1967,) 158-159.

"*seditious,*" he disapproved of the sentence and had the men placed in irons until Brigadier-General Robert Howe of North Carolina arrived to take up the matter.[56]

If that had been all Mercer did, the issue might have ended there, but Colonel Mercer was determined to end the rampant defiance to military authority in Gibson's company, so he took the additional step of disarming the rest of Gibson's riflemen and placed them, under guard, upon fatigue duty. He explained his action in the army's after orders of March 11th:

> *In Consequence of the seditious behavior of some of Capt. Gibson's Comp'y, two of them were Confin'd and a General Court Martial of the Line instituted, the Sentence of which the Commanding Officer Totally Disapproved. It is ordered that the Two Prisoners be Laid in Irons…till the arrival of Genl. Howe. The Rest of the Comp'y, non-com'd officers and Rank and File, having* [displayed] *on* [many] *occasions the Same Seditious and Mutinous Spirit, Shall be Stript of their arms and ammunition, and Put upon Duty of Fatigue under the Direction of the Quarter-Master Genl, who shall be Supported in the Execution of his Duty by a Captain's Guard, Properly Furnish'd with ammunition. The Captain of that Guard is to have orders, that if any of these Seditious and Mutinous Soldiers Shall dare to Refuse to Perform the Duty which the Quarter-Master Shall direct, such offenders shall be put in Irons…*[and] *if any obstruction arise from the same Mutinous Disposition, the Guard is*

---

[56] Brock, ed., "Orderly Book of Capt. George Stubblefield, After Orders, March 11, 1776," *Miscellaneous Papers…in the Collections of the Virginia Historical Society*, 150-151.

to fire on the offenders With such Effect as to kill them if possible.[57]

Mercer's harsh directive was successful in restraining Captain Gibson's company, but his threat to shoot any offender of his orders was challenged by the officers of the unruly company as too extreme, and they appealed to General Howe when he arrived in Williamsburg in mid-March. The result was an apology from Colonel Mercer posted in the general orders of March 17th:

> *Head-Quarters, March 17, 1776*
> *General Orders -- Col. Mercer, sensible that he exceeded the line of duty in his treatment of capt. Gibson's company, has requested the commanding-officer to declare, in orders, that he had no personal intention in any thing he did, and in this publick manner desires to acknowledge he was wrong, and assures the company he is sorry for what happened. The commanding officer is of opinion, that this officer-like acknowledgement of the colonel's ought to be satisfactory to the company.*[58]

Lieutenant Lynn, disturbed by negative reports and rumors about the conduct of his company that circulated after the incident, sent a copy of Mercer's apology to the gazettes, which published it in early April.[59] An angry Colonel Mercer, who had returned to northern Virginia to command the 3rd Virginia Regiment, replied to Lynn with his own letter to the gazettes.

---

[57] Ibid.
[58] Purdie, "April 5, 1776," *Virginia Gazette*, 3.
[59] Ibid.

Fredericksburg, April 10, 1776

The publick, to whom lieut. Lynn thought proper to report an affair of military discipline, will naturally conclude from his publication that I have injured the characters of the men of capt. Gibson's company of regulars. I aimed at mending the character of that company, and hope I have not missed my aim. In attempting this necessary service, it is true, some deviation was made from the line of duty; but whose deviation from duty was most injurious to that company, and to the cause in which we are engaged, I beg leave also to submit to the publick: That of an officer who quells a mutinous spirit in the troops, or of those officers who, by a neglect of discipline, had, after some months training, obliged me to take the trouble of reducing their men to some degree of military order.[60]

Colonel Mercer asserted that his actions towards Gibson's company were for the good of both Gibson's men (who lacked sufficient discipline) and the army in general. He also placed the blame for the incident upon Lieutenant Lynn and his fellow officers who, Mercer contended, had neglected to adequately discipline their troops in the months leading up to the incident and had thus helped create the mutinous spirit in the men.

The arrival and brief stay of Brigadier General Robert Howe allowed Colonel Mercer to leave Williamsburg and join his regiment in northern Virginia. General Howe continued Colonel Mercer's emphasis on discipline and drill, but departed

---

[60] Purdie, "April 19, 1776," *Virginia Gazette*, 4.

within a week of his arrival for North Carolina.[61] Colonel Mordecai Buckner of the 6th Regiment and Lieutenant Colonel Thomas Bullett, the Adjutant General, assumed command of the troops in Williamsburg until Major General Charles Lee arrived at the end of March.

---

[61] Brock, ed., "Orderly Book of Capt. George Stubblefield, March 19, 1776," *Miscellaneous Papers...in the Collections of the Virginia Historical Society*, 155-156.

Chapter Four

Preparation for War Increases

April – August 1776

Major-General Lee had been ordered to Williamsburg by the Continental Congress, who was concerned by reports of a British expedition against the southern colonies in 1776. Lee, a former British officer with extensive military service in Europe, served with General Washington in Boston in 1775. Although he was a native of Britain and had only arrived in the colonies in 1773, Lee earned the trust and admiration of Congress, many of whose members viewed him as the most militarily knowledgeable officer in the army.

British attention to the southern colonies in the spring of 1776 not only caused the Continental Congress to send General Lee south to take command of the region's defense, but also convinced Congress in late March to place the three remaining Virginia regiments of regulars, the 7th, 8th, and 9th Regiments, as well as two additional regiments from South Carolina, onto continental establishment.[1]

Congress also authorized the formation of a continental artillery company in Virginia, instructing General Lee to, "*set on foot the raising of a company of artillery,*" under the command of Captain Dohicky Arundel, one of the first of many

---

[1] Ford, ed., "Proceedings of the Continental Congress, March 25, 1776," *Journals of the Continental Congress*, Vol. 4, 235.

French volunteers who offered their service to the American cause.[2] Congress left it to Virginia's leaders to appoint the remaining officers of the new continental artillery company.

What the Congress was apparently unaware of, however, was that the 4th Virginia Convention had authorized the formation of a company of artillery two months earlier on January 10, 1776 to consist of one captain, three lieutenants, one sergeant, four bombardiers, eight gunners and 48 matrosses.[3] A month after that, on February 13th, the Virginia Committee of Safety appointed the officers for the artillery company. James Innes was appointed captain, Charles Harrison 1st lieutenant, Edward Carrington 2nd lieutenant and Samuel Denny 3rd lieutenant of the company.[4]

Congress's appointment of Captain Arundel to command a continental artillery company in Virginia caused some confusion among Virginia's leaders, who were uncertain where Arundel's appointment left their appointee, Captain James Innes. General Lee attempted to resolve the matter soon after he arrived in Virginia by appointing Innes, who Lee noted was, "*a man of great zeal, capacity, and merit*," but who, "*professes himself utterly ignorant*," of the artillery service, to serve as major of the 9th Regiment in place of the recently deceased Matthew Donavan.[5] Although Congress never formally

---

[2] Ford, ed., "Proceedings of the Continental Congress, March 19, 1776," *Journals of the Continental Congress*, Vol. 4, 212.

[3] Scribner and Tarter, eds., "Proceedings of the 4th Virginia Convention, January 10, 1776," *Revolutionary Virginia, Road to Independence*, Vol. 5, 378.

[4] Scribner and Tarter, eds., "Proceedings of the Virginia Committee of Safety, February 13, 1776," *Revolutionary Virginia, Road to Independence*, Vol. 6, 91.

[5] ----- "General Lee to the President of Congress, April 19, 1776," *The Lee Papers*, Vol. 1, 434.

authorized Innes's promotion by Lee, Innes did receive a continental commission as major in the 5th Virginia regiment before the summer ended. Congress also added two lieutenants and forty additional men to the artillery company, effectively creating two companies of continental artillery in Virginia.[6]

General Lee's arrival in Williamsburg in late March prompted several changes to Virginia's military arrangements. Critical of the scattered deployment of the regiments, which left the capital vulnerable to attack, (something Lee believed was very likely with a large British force gathering off of North Carolina's coast) the American commander ordered the 3rd and 5th Virginia Regiments to march to Williamsburg.[7] He also ordered Colonel Christian of the 1st Virginia to thoroughly reconnoiter the north bank of the James River to identify defensive strongpoints in expectation of an enemy landing.[8]

General Lee expressed his concerns about a possible British attack to General Washington a week after his arrival.

> *My situation is just as I expected. I am afraid that I shall make a shabby figure, without any real demerits of my own. I am like a dog in a dancing school – I*

---

[6] Ford, ed., "Proceedings of the Continental Congress, May 18, 1776," *Journals of the Continental Congress*, Vol. 4, 364.
Note: When Congress authorized the formation of an artillery regiment in Virginia on November 26, 1776, it specified that, "the two [artillery] companies, already raised [in Virginia] were to be incorporated into the new artillery regiment. (See: Ford, ed., "Proceedings of the Continental Congress, November 26, 1776," *Journals of the Continental Congress*, Vol. 6, 981).
[7] --- "General Lee to Colonels Peachy and Mercer, April 2, 1776," *The Lee Papers*, Vol. 1, 369.
[8] --- "General Lee to Colonel Christian, April 3, 1776," *The Lee Papers*, Vol. 1, 370.

> *know not where to turn myself. The circumstances of the country, intersected by navigable rivers; the uncertainty of the enemy's designs and motions, who can fly in an instant to any spot where they chose with their canvas wings* [ships], *throw me or would have thrown Julius Caesar, into this inevitable dilemma.*[9]

Lee included a brief assessment of Virginia's forces and capability.

> *The regiments in general are very complete in number, the men (those that I have seen) fine; but a most horrid deficiency of arms – no intrenching tools – no guns* [cannon]…*meant for service.*[10]

General Lee expressed similar concerns a week later in a letter to Richard Henry Lee, one of Virginia's delegates to Congress. Lee reported that he was experimenting with spears to offset the severe shortage of arms for the troops. He believed he was making progress in, *"conciliating your soldiers to the use of spears,"* and described how he employed them in a recent exercise.

> *We had a battalion out this day; two companies of the strongest and tallest were armed with* [spears]; *they were formed something like the Triarii of the Romans, in the rear of the battalions…. It has a fine effect to the*

---

[9] --- "General Lee to General Washington, April 5, 1776," *The Lee Papers*, Vol. 1, 376-377.
[10] Ibid., 377.

eye, and the men in general seemed convinced of the utility of the arrangement.[11]

Lee's concern of a looming British attack led him to propose to the Committee of Safety, with the full agreement of an officers' council, the removal of the inhabitants of Norfolk and Princess Anne County who were within supporting distance of Lord Dunmore's floating town in the Elizabeth River between Norfolk and Portsmouth.[12] The committee unanimously agreed to Lee's proposal and ordered the removal of all inhabitants and their livestock located between Norfolk and Great Bridge and Kemp's Landing.[13] The regiments posted in Suffolk, namely the 2nd, 4th, and 8th Regiments, were tasked by General Lee with overseeing the forced evacuation.

The Committee of Safety also addressed General Lee's redeployment of regiments, explaining their rational for originally dispersing them and suggesting that redeploying the units to Williamsburg would significantly complicate the provisioning of the troops. News from Philadelphia that Congress had accepted Virginia's remaining provincial regiments, the 7th, 8th, and 9th Virginia, onto continental establishment prompted General Lee to rescind his orders for the 3rd Virginia to march to Williamsburg. He explained the change of plans to Colonel Mercer.

---

[11] --- "General Lee to Richard Henry Lee, April 12, 1776," *The Lee Papers*, Vol. 1, 416-417.
[12] Scribner and Tarter, eds., "Major General Charles Lee to the President of the Committee of Safety of Virginia, April 8, 1776," *Revolutionary Virginia, Road to Independence*, Vol. 6, 352.
[13] Scribner and Tarter, eds., "Unanimous Resolution for the Evacuation of Parts of Norfolk and Princess Anne Counties, April 10, 1776," *Revolutionary Virginia, Road to Independence*, Vol. 6, 369-370.

> *When I order'd your Regiment to march down to Williamsburg, it was not my intention to strip the Northern Neck of the Troops necessary for its defence; I proposed to replace your Regiment with Col. Dangerfield's [7th Virginia] then in the Provincial Pay; This exchange was founded on a very solid reason, for had it been necessary to march a Body out of the Province the Continental Forces alone were at my disposal – but since my Letter the whole have been put on the Continental Establishment.*[14]

This meant that General Lee now had the authority to send any of Virginia's nine continental regiments anywhere he wished. With the 7th Virginia Regiment closer to Williamsburg, Lee ordered half of that unit to the capital (leaving five companies in Gloucester at the request of the county).[15]

Colonel William Peachy's 5th Virginia Regiment did not receive a change of orders and arrived in Williamsburg around the time Lee ordered Mercer to remain in the Northern Neck. With reports from North Carolina suggesting that General Lee might need to march Virginia reinforcements there to assist, the 5th Regiment was ordered across the James River to join the 2nd, 4th, and 8th Regiments in the vicinity of Suffolk.[16]

---

[14] --- "General Lee to Colonel Mercer, April 10, 1776," *The Lee Papers*, Vol. 1, 409.

[15] Scribner and Tarter, eds., "Gloucester County Committee to the Virginia Committee of Safety, April 22, 1776," *Revolutionary Virginia: Road to Independence*, Vol. 6, 441.

[16] Brock, ed., "Orderly Book of Capt. George Stubblefield, April 13, 14, 21, 1776," *Miscellaneous Papers… in the Collections of the Virginia Historical Society*, 162-163.

General Lee joined these forces in Suffolk on April 21st, leaving Brigadier General Andrew Lewis, who had recently arrived in the capital, in charge of the forces in Williamsburg, namely the 6th Virginia, part of the 1st Virginia, and by early May half of the 7th Virginia. While Lee supervised the evacuation of large portions of Norfolk and Princess Anne County, he received alarming letters from North Carolina that convinced him that the British intended to, *"attack these Provinces with a very considerable force."*[17]

Although still uncertain of where they would attack, Lee had nearly half of Virginia's continental troops (the 2nd, 4th, 5th, and 8th Virginia Regiments) on the south side of the James River ready to respond to either a British attack in southeastern Virginia or further south in the Carolinas. General Lee was not concerned about Lord Dunmore's small force, he focused on the growing British fleet of warships anchored off the Cape Fear River in North Carolina.

Williamsburg was protected by the 6th Regiment as well as most of the 1st Regiment and half of the 7th Regiment. Several companies of the 1st Virginia were also posted in Hampton and five companies of the 7th Virginia remained in Gloucester County, across the York River from Williamsburg. The 3rd Virginia remained in Prince William and Fairfax Counties guarding the Potomac and the 9th Virginia remained on the Eastern Shore. All waited with their commanding officer, General Lee, for the British to make a decisive move.

Lee returned to Williamsburg in early May, leaving Colonel William Woodford of the 2nd Regiment in command at Suffolk. The arrival of several British transport ships loaded

---

[17] --- "General Lee to the President of Congress, April 24, 1776," *The Lee Papers*, Vol. 1, 449.

with troops off the Cape Fear River in North Carolina convinced Lee that Wilmington was most likely the British objective, so he decided it was time to march south to lead the defense there.

General Lee ordered only one Virginia regiment, the 8th Virginia, which he considered, *"the strongest battalion we have,"* to march south with him in mid-May.[18] It appears that Lee initially regretted his choice of the 8th Regiment for in a letter to the Virginia Committee of Safety written in Halifax, North Carolina just a week after they left Virginia, Lee berated the regiment for its unmilitary behavior.

> *The disorderly mutinous and dangerous disposition of the soldiers of the 8th Regiment have detain'd me longer in this place than I cou'd have wish'd, more particularly as we hear (tho the accounts are not well authenticated) that the whole fleet of Transports under Lord Cornwallis is arrived at Cape Fear. We have at length after infinite trouble got this Banditti out of the Town, and of course I set out myself immediately.*[19]

Lee complained that desertion among the troops was rampant and he urged Virginia's leaders to take action to halt it.

> *The spirit of desertion in these back Country troops is so alarmingly great, that I must submit it to the wisdom of the convention, whether it is not of the utmost importance to devise some means to put a stop to it,*

---

[18] --- "General Lee to the President of Congress, May 7, 1776," *The Lee Papers*, Vol. 1, 480.

[19] Tarter, ed., "General Charles Lee to Edmund Pendleton, May 24, 1776," *Revolutionary Virginia: The Road to Independence,* Vol. 7, Part One, 248-249.

> *before it spreads, by enjoining the Committees of the different Counties to seize every Soldier, who cannot produce an authenticated discharge or pass....*[20]

He confessed that Virginia's civil authorities were better judges of what to do to stem the high desertion rate, but warned that

> *I can only affirm that unless some effectual method is devised and adopted, it will be impossible for us to keep the Field. The old Countrymen, particularly the Irish, whom the Officers have injudiciously inlisted in order to fill up their Companies, have much contaminated the Troops; and if more care is not taken on this head, for the future, the whole Army will be one mass of disorder, vice and confusion....*[21]

Although General Lee held the troops of the 8th Virginia in low regard during the march south, he was careful not to include the regiment's officers in his criticism, writing that,

> *Altho I have so great reason to complain of the misconduct of this Regiment, I must do the Officers (particularly the Field Officers) the justice to say, that their conduct is in general very satisfactory.*[22]

About a week after General Lee's letter to the Committee of Safety, he ordered the 8th Virginia to march even further south, to Charleston, South Carolina. Lee correctly deduced that

---

[20] Ibid.
[21] Ibid.
[22] Ibid.

Charleston was the true objective of the British expedition and he rushed ahead to get there in time to prevent its capture.[23]

When the 8th Virginia arrived in Charleston on June 23, the British under General Henry Clinton and General Charles Lord Cornwallis, were well on their way to launching an attack on the outer defenses of the city. The key patriot defensive position protecting Charleston was a fort on Sullivan's Island. It guarded the harbor channel to Charleston with walls constructed of palmetto logs and sand that were sixteen feet thick. The spongy nature of the palmetto logs and the thickness of the walls allowed the fort to absorb even the heaviest cannon blasts of the British navy.[24] A garrison of only 350 South Carolina troops manned the 31 cannon in the fort, but they were supported by a large detachment of 750 troops on the north end of Sullivan Island who were positioned to prevent a British landing from an adjacent island.[25]

Such a landing was exactly what the British intended to do with their 2,200 troops.[26] The plan called for the British navy to pound the rebel fort with cannon fire while the British army crossed a narrow cut of water that separated Sullivan's Island from Long Island (upon which the British army had landed weeks earlier) then march four miles to the fort and capture it by storm.

Colonel Muhlenberg's 8th Virginians were not initially deployed on Sullivan's Island, but rather, at Haddrell's Point

---

[23] --- "General Lee to Edmund Pendleton, June 1, 1776," *The Lee Papers*, Vol. 2, 51.
[24] Edwin C. Bearss, *The Battle of Sullivan's Island and the Capture of Fort Moultrie: A Documented Narrative and Troop Movement Maps*, (U.S. Dept. of the Interior, 1968), 58-59.
[25] Ibid., 57, 59-60.
[26] Ibid., 29, 50.

on the mainland, to guard against a possible British crossing onto the mainland from the outer islands. Thus, when the British navy commenced its attack upon the American fort on Sullivan's Island on June 28th, the 8th Virginians were largely spectators.

The British attack unraveled quickly, however, when General Clinton discovered that he had no way to cross the cut between Sullivan's and Long Island without suffering unacceptably high losses. The narrow channel between the two islands was deeper than Clinton had been informed (seven feet at low tide) and was unfordable, and the British commander believed any attempt to use boats to either cross the cut or land further up the island would lead to enormous casualties among his troops.[27] So with Commodore Peter Parker's warships fully engaged against the fort, the best General Clinton could do was create a distraction by threatening to cross the cut, thereby keeping the 750 rebel infantry at the cut away from their fort.

The outcome of the battle thus depended on the British warships under Commodore Parker and Colonel William Moultrie's garrison inside the fort. Relatively safe behind their thick walls of palmetto logs and sand, the American garrison punished the stationary British warships with 26 pound, 18 pound, 12 pound, and 9 pound ordinance.[28] Several of Commodore Parker's frigates attempted to sail past the fort but became grounded in low water and fell out of action. One of these was eventually abandoned and burned.

---

[27] William J. Morgan, ed., "Comments by Major General Henry Clinton Upon the Naval Attack on Sullivan's Island, June 28, 1776," and "Major General Henry Clinton to Lord George Germain, July 8, 1776," *Naval Documents of the American Revolution*, Vol. 5, (Washington: 1970), 802 and 983.

[28] Bearss, 58-59.

General Lee, who was on the mainland when the attack began, crossed over to Sullivan's Island around 5 p.m. and was pleased to see that Colonel Moultrie had the situation well in hand.[29] Moultrie had lost over a score of men, but he was confident that the damage his men inflicted upon the enemy was far worse.[30] On that count Moultrie was correct, for the American fire did significantly more damage to the British navy than Commodore Parker's fire did to the fort.

General Lee was surprised and impressed at the resolve of the garrison and described the engagement as, *"one of the most furious cannonades I ever heard or saw."*[31] Lee admitted after the battle that,

> *The behavior of the Garrison, both men and officers, with Colonel Moultrie at their head, I confess astonished me; it was brave to the last degree. I had no idea so much coolness and intrepidity could be displayed by a collection of raw recruits, as I was witness of in this garrison.*[32]

Up to this point of the battle, the 8th Virginians had remained on the mainland in reserve. Late in the afternoon, General Lee ordered Colonel Muhlenberg to reinforce the large American detachment defending the cut at the northern end of Sullivan Island.[33] Muhlenberg's men must have displayed some zeal to

---

[29] Ibid., 84-85.
[30] --- "General Lee to the President of the Virginia Convention, June 29, 1776," *The Lee Papers,* Vol. 2, 93.
[31] Ibid.
[32] Ibid.
[33] Bearss, 86.

join the fight for General Lee paid them a compliment in his report to the Virginia Convention after the battle:

> *I know not which Corps I have the greatest reason to be pleased with, Muhlenberg's Virginians, or the North Carolina troops. – they are both equally alert, zealous and spirited.*[34]

Muhlenberg's Virginians may have been spirited and eager to join the fight, but it appears they saw little actual action on Sullivan's Island. The honor of the day belonged to Colonel Moultrie's garrison, who withstood the British navy's intense bombardment and returned fire with such deadly effect that Commodore Parker eventually disengaged and withdrew. The British threat to Charleston thus ended in defeat for the attackers, who suffered significant damage to their ships and heavy losses among their crew.

The damage to the British navy was so great that they did not make another attempt against Charlestown. After three weeks of repair to their ships, the British sailed away from Charleston and the South.[35] It was time to join General William Howe in New York.

Gwynn's Island

When General Lee left Williamsburg in mid-May, the focus of nearly everyone in the capital was on the proceedings of the 5th Virginia Convention. On May 15th, the Convention took the enormous step of approving, unanimously, a resolution

---

[34] --- "General Lee to the President of the Virginia Convention, June 29, 1776," *The Lee Papers,* Vol. 2, 93.
[35] Bearss, 102-105.

for independence. The troops then in Williamsburg, primarily the 6th Regiment, most of the 1st Regiment and half of the 7th Regiment, were paraded in Waller's Grove near the Capitol the following day to hear the adopted resolution read aloud. Cannon and musket fire saluted the resolution and several toasts were offered to American independence, the Continental Congress, and General Washington. In the evening the jubilant soldiers and inhabitants of Williamsburg enjoyed an evening of illuminations, music and drink.[36]

Ten days after the celebration, on the morning of May 26th, a detachment of Captain Thomas Posey's company of 7th Virginians posted at Burton Point in Gloucester County (near the mouth of the Piankatank River), spotted Lord Dunmore's fleet, which had been largely inactive over the winter and spring, off of Gwynn's Island. They rushed word to Colonel Daingerfield at Gloucester Courthouse who alerted the local militia and led the several companies of his regiment then in Gloucester, towards Gwynn's Island.[37]

Captain Posey, whose company was posted in several spots near Gwynn's Island, also raced there and found, *"the militia assembled, which appear'd to be in the utmost consternation, some running one way, and some another, under no kind of control or regularity."*[38]

Colonel Daingerfield ordered both his troops and the militia to advance closer to the shore to prevent the enemy, which had anchored off of Gwynn's Island, from landing on the mainland. Captain Posey observed that,

---

[36] Purdie, "May 17, 1776," *Virginia Gazette*, 3.
[37] Purdie, "May, 26 1776," *Virginia Gazette*, 3.
[38] Thomas Posey, "May 27, 1776," *Revolutionary War Journal of Thomas Posey, Jan. 12, 1776 – May 24, 1777*, Indiana Historical Society. Library, (Unpublished). Henceforth referred to as Posey Journal.

> *The whole were put in motion, (though I must confess the militia were in very great motion before the orders were given). However, these orders served to put them in something grator; for as soon as we came neare enough for the grape[shot], and cannon shot to whistle over our heads, numbers of the militia put themselves in much quicker motion, and never stopped...to look behind them until they had made the best of there way home.*[39]

The militia's conduct was not the only behavior Captain Posey found fault with. He was also critical of the regular troops:

> *I cant say that our regulars deserved any great degree of credit for after two or three getting a little blood drawn, they began to skulk and fall flat upon there faces.*[40]

Despite their apprehension, the 7th Virginians and at least some of the militia, held their ground and endured cannon fire and heavy rain all evening. As the hours passed, they grew more determined to face the enemy. Captain Posey recalled, "*We began to grow very firm and only wish them to come into the bushes, where we are certain of beating them.*"[41]

Rather than attack the mainland, however, Lord Dunmore contented himself with fortifying Gwynn's Island. A channel of water over a hundred yards wide separated the two sides. Guarded by cannon aboard several ships as well as on the

---

[39] Ibid.
[40] Ibid.
[41] Ibid.

island, Governor Dunmore deemed the island secure and erected earthworks and an encampment on the neck closest to the mainland. British cannon fire frequently rang out over the course of several weeks, answered by occasional rifle fire from the mainland.

Although rebel fire had a minimal impact on Dunmore's force, the royal governor's men suffered greatly from disease. In a letter to Lord Germain in London, Dunmore described the impact illness had on his troops:

> *I am extreamly sorry to inform your Lordship that the Fever of which I informed you in my Letter No. 1, has proved a very Malignant one and has carried off an incredible Number of our People, especially the Blacks, had it not been for this horrid disorder, I am Satisfied I should have had two thousand Blacks, with whom I should have no doubt of penetrating into the heart of the Colony...There was not a ship in the fleet that did not throw one, two, or three or more dead overboard every night.*[42]

The Virginians were well aware of Dunmore's losses. General Andrew Lewis in Williamsburg forwarded reports of daily sightings of bodies from Gwynn's Island to General Lee. "*A Great Mortality among the Enemy, some both white and black, are discovered floating every day*"[43] Although

---

[42] William Clark, ed., "Lord Dunmore to Lord Germain, June 26, 1776," *Naval Documents of the American Revolution,* Vol. 5, (Washington: 1970), 756.

[43] Clark, ed., "Brigadier General Andrew Lewis to Major General Charles Lee, June 12, 1776," *Naval Documents of the American Revolution,* Vol. 5, 501.

Dunmore's force suffered terribly from smallpox and fever, they were still relieved to be on land. There was fresh provision and they were relatively secure from attack.[44]

The Virginians on the mainland, both continental and militia, worked hard to change this. Despite frequent enemy gunfire, steady progress on earthworks and artillery placements along the shore facing Dunmore's encampment occurred in June. General Lewis informed General Lee of developments:

> *I have ordered several Pieces of Cannon at Gloucester Town to be mounted, which the workmen are about, in order to have them mounted opposite the Enemy and if possible, to prevent some small armed Vessels getting out which lie between the mainland and the Island. I have sent under the Command of Col. Mercer three companies to reinforce Col. Dangerfield's Battalion....*[45]

Colonel Mercer's 3rd Virginia Regiment arrived in Williamsburg from its post in northern Virginia sometime in early June to reinforce the 1st, 2nd, (who had been ordered to Williamsburg upon Dunmore's departure from Norfolk) and 6th Regiments in and around Williamsburg. Mercer's stay in the capital was brief; Congress promoted him to brigadier-general in early June and ordered him to proceed north to take command of new troops with General Washington's army in New York.[46]

---

[44] Clark, ed.," Lord Dunmore to Lord Germain, June 26, 1776", *Naval Documents of the American Revolution,* Vol. 5, 756.

[45] --- "Brigadier General Lewis to Major General Lee, June 12, 1776", *The Lee Papers,* Vol. 1, 63.

[46] Force, ed., "Mercer to Congress, June 15, 1776," *American Archives, Fourth Series*, Vol. 6, 903.

Skirmishing continued at Gwynn's Island throughout June while the troops from the 3rd and 7th Virginia Regiments, a company of artillerists, and local militia worked mightily to prepare artillery positions and haul cannon into place. In one ship to shore skirmish, the Virginians were able to seize a small vessel loaded with rum and other spirits that had drifted too close to shore. General Lewis described the attack in a letter to General Lee:

> *Our men took a small sloop endeavouring to get out of the Narrows between the Island and our breastwork. She having run a Ground, a few men in two small Canoes boarded her, five men who were all her crew endeavoured to escape by swimming – three of which were shot from the shoar and sunk. Two hogsheads of Brandy ½ Ditto of Rum, some tools and ropes…were taken out for the use of our Troops there, who were in need of the brandy and rum, as the water is very bad.*[47]

The British intensified their bombardment of the Virginians in the latter half of June. Besides slowing the work of the fatigue parties, the cannon fire did little damage.

The situation changed on the morning of July 9th, when the Virginians completed their artillery batteries consisting of two 18 pound cannon and an experimental wooden mortar overlooking the narrow channel between the island and mainland, and a second battery of five 6 and 9 pound cannon positioned two hundred yards to the east of the 18 pounders. Both batteries revealed themselves to the enemy with a sudden

---

[47] --- "General Lewis to General Lee, June 12, 1776," *The Lee Papers*, Vol. 2, 64.

barrage that shocked Lord Dunmore and his forces. Captain Posey described what happened in his diary:

> *At ten o clock orders were given by Genl. Lewis to open the whole of the batteries, two of which were opposite the enemies encampment and the main battery within a few hundred yards of the fleet. The fireing was kept up in a very regular manner from the whole of our works for near two hours; during which time they received great damage....*[48]

General Lewis reportedly aimed and fired the first cannon at Lord Dunmore's ship, the *Dunmore*. The eighteen pound ball crashed through its stern, killing an aide and destroying Dunmore's cabin. More cannonballs landed amidst the British camp and fleet, creating panic and an almost immediate evacuation of the island.[49]

At some point in the bombardment, Captain Arundel attempted to fire a wooden mortar of his own design. Warned by several officers including General Lewis that it was too dangerous to attempt, Arundel loaded and fired the mortar, which blew up, killing him instantly.[50]

Lord Dunmore's troops on Gwynn's Island, stunned and panicked by the unexpected American bombardment, scrambled to evacuate the island while troops from the 7th Virginia attempted to procure boats to cross the channel and

---

[48] Posey Journal, July 9, 1776.
[49] Peter Wrike, *The Governor's Island,* (Gwynn, VA: The Gwynn's Island Museum, 1993), 79.
[50] Wrike, 81.

press the enemy. Captain Posey described the scene in his journal.

> *Upon the enemies receiving this very unexpected stroke, they gave immediate orders to evacuate the Island. On the discovery of which, orders were given to cross into the island and endeavor to harass the enemy in the rear. Col. McClannahan was directed to take command of about 200 men for the afore said purpose.*[51]

A lack of boats, however, delayed the American crossing until the next morning. By then, Dunmore and his men had evacuated the island and returned to their ships. Captain Posey was one of the first Virginians on the island and described the landing in his diary:

> *Crossed into the Island but no fighting ensued except a few shot. By one oclock the whole of the enemy had evacuated and embarked...I cannot help observeing, that I never saw more distress in my life, than what I found among some of the poor deluded Negroes which they could not take time, or did not chuse to cary off with them, they being sick. Those that I saw, some were dying, and many calling out for help; and throughout the whole Island we found them stre'd about, many of them torn to pieces by wild beasts – great numbers of the bodies having never been buried.*[52]

---

[51] Posey Journal, July 9, 1776.
[52] Posey Journal, July 10, 1776.

British losses at Gwynn's Island are difficult to ascertain. Captain Posey estimated, *"that at least 4 or 500 negroes lost their lives."*[53] Posey added that another 150 [white] soldiers were also lost.[54] The vast majority of these deaths occurred prior to the attack as a result of disease. Such losses significantly hampered the effectiveness of Dunmore's force and explain his feeble response to the attack.

The events at Gwynn's Island exasperated Lord Dunmore. His men were weak from illness and demoralized by defeat and there was little hope of assistance from Britain. By mid-August Dunmore had had enough and abandoned his efforts to regain Virginia. Half of his force sailed to St. Augustine, Florida and the other half sailed, with him, to New York.

Most of the 7th Virginia Regiment returned to Gloucester Courthouse a few days after the battle and those of the 3rd Virginia rejoined the rest of their regiment in Williamsburg. A small guard was left on Gwynn's Island to keep watch for the enemy.[55] A week later Colonel Daingerfield resigned his commission, leaving Lieutenant Colonel McClanahan in command.[56]

News of the Declaration of Independence reached Williamsburg on July 24 and the next day Virginia's new state government had the declaration read at the Capitol, Courthouse and Palace, *"amidst the acclamations of the people."*[57] Troops

---

[53] Ibid.
[54] Ibid.
[55] Posey Journal, July 11, 1776.
[56] Posey Journal, July 17, 1776.
[57] Purdie, "July 26, 1776," *Virginia Gazette*, 2.

from the 1st, 2nd, 3rd, and 6th Virginia Regiments paraded in Williamsburg and fired a number of volleys in celebration.[58]

To the south, the departure of the British from Charleston caused General Lee to turn his attention to neighboring Georgia at the end of July. Officials there had expressed concerns about their defenseless state and appealed for assistance against the British navy, Indians allied with the British, and from British troops operating from St. Augustine and several smaller outposts in East Florida.[59]

General Lee resolved to march to Georgia with a force that included the 8th Virginians. Colonel Muhlenberg's regiment, much reduced by the rigors of service in the Carolinas, had over 150 men unfit for duty due to illness.[60] Major Peter Helpenstine became so ill he resigned his commission in early August and returned to Virginia, but died soon after.[61] General Lee appointed Richard Campbell acting major of the 8th Virginia on August 10th, (pending the approval of Congress).[62]

Lee was eager to strike the British in East Florida to put a stop to their harassment of Georgia. This largely wilderness region between Georgia and the British stronghold at St.

---

[58] Ibid. and Charles Campbell, ed., "General Orders for July 19, 21, and August 2, 1776," *The Orderly Book of that Portion of the American Army Stationed at or Near Williamsburg...March 18th, 1776 to August 28th, 1776*, (Richmond, VA, 1860), 63-63, 67.

[59] ----- "General Lee to President Rutledge, July 22, 1776," *The Lee Papers,* Vol. 2, 159.

[60] Charles H. Lesser, ed., "Monthly Return of the Forces in South Carolina, July 1776," *The Sinews of Independence: Monthly Strength Reports of the Continental Army*, (University of Chicago Press, 1976), 27.

[61] Francis, B. Heitman, *Historical Register of Officers of the Continental Army During the War of the Revolution, April 1775 to December 1783*, (Washington, D.C., 1914), 284.

[62] Ford, ed., "Resolution of Congress, January 21, 1777," *Journals of the Continental Congress,* Vol. 7, 52.

Augustine held several British outposts from which incursions into Georgia had been launched. The losses suffered by Georgians from these British raids involved mostly slaves and cattle, but the potential for greater loss was high, so General Lee resolved to break up the British posts in East Florida.[63]

Supply and transport delays pushed General Lee's march to Georgia into mid-August.[64] The 8th Virginia, with approximately 300 men fit for duty, made up part of Lee's force and soon after they reached Savannah, Lee ordered the Virginians further south another 30 miles to the settlement of Sunbury.[65] Little of note occurred on the march, and the 8th Virginia was recalled to Savannah where they learned that Congress had ordered General Lee to Philadelphia, effectively ending his expedition to Georgia.[66]

A few weeks earlier, General Lee had issued orders to send Colonel Muhlenberg's sick and unfit Virginians who had remained in Charleston (approximately 150 men) back to Virginia as soon as they recovered.[67] Colonel Muhlenberg and an unknown number of his men also marched north upon General Lee's departure, leaving Major Campbell and the

---

[63] ----- "General Lee to Richard Peters, August 2, 1776," *The Lee Papers*, Vol. 2, 188-189.

[64] ----- "Orders Issued on the Expedition to Georgia, August 12 and 16, 1776," *The Lee Papers*, Vol. 2, 251-252.

[65] Lesser, ed., "Monthly Return of the Forces in South Carolina, July 1776," *The Sinews of Independence: Monthly Strength Reports of the Continental Army*, 27 and "Orders Issued on the Expedition to Georgia, August 21, 1776," *The Lee Papers*, Vol. 2, 253.

[66] Ford., ed., John Hancock to General Lee, August 8, 1776," *Journals of the Continental Congress*, Vol. 5, 639, and "Orders, September 9, 1776," *The Lee Papers*, Vol. 2, 258-259.

[67] -----"General Lee to General Armstrong, August 15, 1776," *The Lee Papers*, Vol. 2, 230.

remainder of the 8th Virginia behind in Charleston.[68] They eventually marched back to Virginia in January. Colonel Muhlenberg reported their arrival to General Washington in late February.

> *The detachment from the southward arrived here this week in a shattered condition, having only seventy men fit for duty; so that it will be almost impossible to march the men so soon as I could wish.*[69]

Much transpired in the fall and winter of 1776 while the 8th Virginia languished in South Carolina. The focus of the war shifted north to New York in August, where General William Howe amassed an enormous invasion force of 30,000 British and Hessian troops on Staten Island to capture New York and crush the American rebellion. With the threat to Virginia significantly diminished by Dunmore's departure, the Old Dominion was free to send assistance northward to support General Washington and the American army in New York. The support of the Virginia continental line proved crucial to the American cause in the coming months.

---

[68] Henry A. Muhlenberg, "Colonel Muhlenberg to his Father, December 20, 1776," *The Life of Major-General Peter Muhlenberg of the Revolutionary Army*, (Philadelphia: Cary and Hart, 1849), 69
and Ford, ed., "Resolution of Congress, January 21, 1777," *Journals of the Continental Congress,* Vol. 7, 52.

[69] Frank E. Grizzard, ed., "Colonel Muhlenberg to General Washington, February 23, 1777," *The Papers of George Washington, Revolutionary War Series,* Vol. 8, (Charlottesville: University Press of Virginia, 1998), 428.

Gwynn's Island

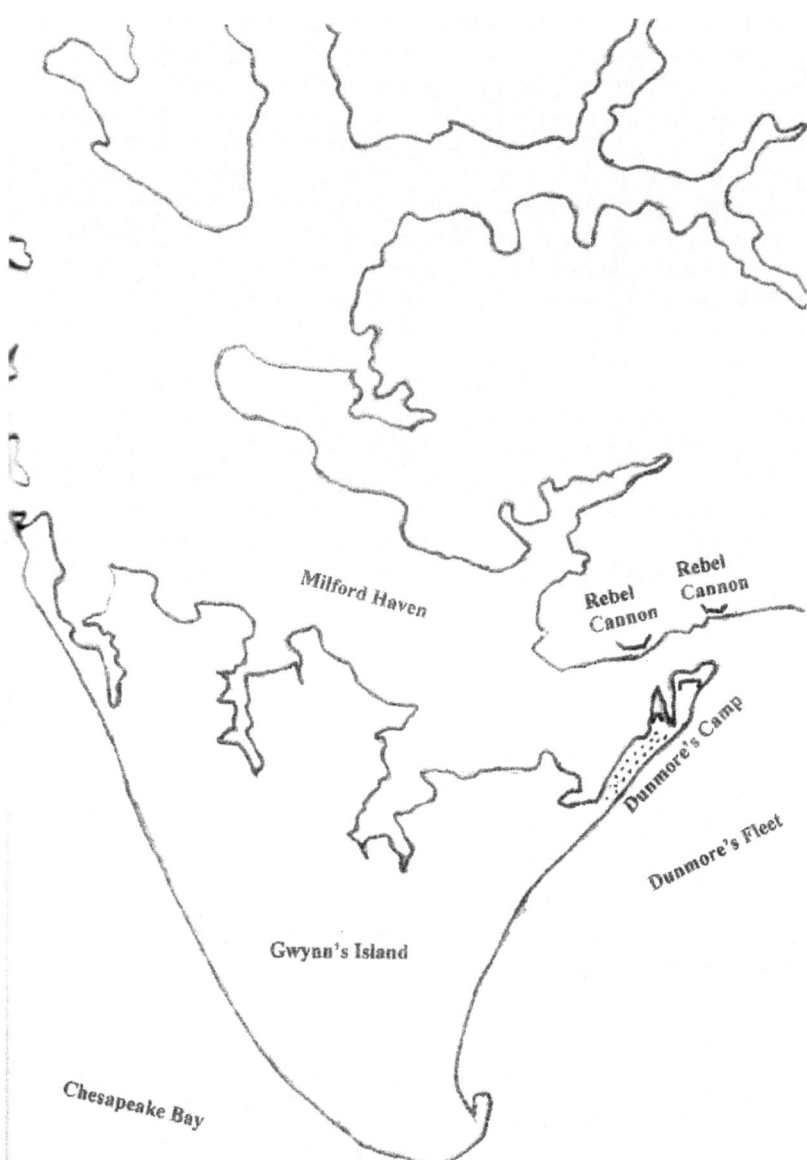

Chapter Five

Virginia Regiments Join Washington's Army

August -- November 1776

On July 20, 1776 the Continental Congress, concerned about the enormous British force that had encamped on Staten Island weeks earlier, called upon Virginia to send two continental regiments northward to reinforce the recently formed Flying Camp under General Hugh Mercer in New Jersey.[1] This detachment was originally intended to be 10,000 strong, made up of militia from Pennsylvania, Maryland and Delaware.[2] It was to be a sort of ready reserve for General Washington, posted in New Jersey to defend that state, but ready to support Washington in New York if needed. Concern at the slow progress of raising the force likely caused Congress to request the two Virginia regiments.

Word of the request reached Williamsburg in early August. General Lewis acknowledged that the 1st and 2nd Virginia Regiments were entitled to the honor of marching north, but with the enlistments of these troops about to expire, it was necessary to re-enlist the soldiers before they marched.[3] Efforts

---

[1] Ford, "Proceedings of the Continental Congress, July 20, 1776," *Journals of the Continental Congress*, Vol. 5, 597.
[2] Ford, "Proceedings of the Continental Congress, June 3, 1776," *Journals of the Continental Congress*, Vol. 5, 412.
[3] Campbell, ed., "General Orders August 4, 1776," *The Orderly Book of that Portion of the American Army Stationed at or Near Williamsburg...March 18th, 1776 to August 28th*, 68.

to do so began immediately and were largely successful in the 1st Regiment, but most of the troops in the 2nd Regiment balked at Colonel Woodford's appeal to re-enlist. An observer of the recruitment effort noted that the 2nd Regiment,

> *Resisted* [Woodford's] *eloquent harangue at their head, and silently rejected the intended honour he proposed doing them by delaying his resignation that he might lead them on the Field of Glory. They say they will follow Col.* [Charles] *Scott, but he is ordered to the 5th* [Regiment]....[4]

The 3rd Virginia Regiment under newly promoted Colonel George Weedon marched north in mid-August in place of the 2nd Regiment. Weedon's men had enlisted for two years just a few months earlier and were in Westmoreland County on the Potomac River guarding against a potential return of Lord Dunmore when they received orders to march, so they set out a few days before the 1st Virginia with an eighty mile head start. Thomas Marshall, promoted to Lieutenant Colonel, and Andrew Leitch, promoted to Major, assisted Colonel Weedon in command of the 3rd Virginia.[5]

Colonel Isaac Read, who was transferred to the 1st Regiment from the 4th Virginia upon the resignation of Colonel William Christian (who returned to the frontier to participate in an expedition against the Cherokee Indians), led the 1st Virginians northward from Williamsburg. Lieutenant Colonel

---

[4] Lyon G. Tyler, ed., "Gabriel Johnston to Leven Powell, August 6, 1776," *Tyler's Quarterly Historical and Genealogical Magazine*, Vol. 12, (Richmond, VA: Richmond Press, Inc., 1931), 90.
[5] Ford, "Proceedings of the Continental Congress, August 13, 1776," *Journals of the Continental Congress*, Vol. 5, 649.

Francis Epps and Major John Green assisted Colonel Read in command of the 1st Virginia.

General Washington and his army in New York suffered a staggering defeat to the British on Long Island prior to the arrival of the Virginia continentals. Outmaneuvered, Washington's army broke up and fled in panic to the heights of Brooklyn in the engagement. Granted a reprieve by British General William Howe, who failed to aggressively follow his initial victory with a final push that would have likely destroyed a large part of the American army, Washington deftly evacuated his troops to Manhattan under the nose of the British navy and army.

The Virginia continentals did not participate in the fighting at Long Island. When the 3rd Virginia reached General Mercer's Flying Camp in New Jersey it was ordered to continue on to New York. Colonel Weedon rode ahead of his regiment and reached Washington's headquarters on September 11th.[6] Weedon's troops arrived two days later, crossing onto Manhattan via King's Bridge at the northern tip of the island.[7]

Captain John Chilton of the 3rd Virginia recalled that the arrival of the Virginians was met with much enthusiasm from the continental army.

Our Regiment have reached this place in good spirits and generally speaking healthy, tho not quite full,

---

[6] Philander D. Chase and Frank E. Grizzard, Jr., eds. "General Washington to General Mercer, September 11, 1776," *The Papers of George Washington, Revolutionary War Series,* Vol. 6, (Charlottesville: University Press of Virginia, 1994), 285.

[7] Lyon Tyler, ed., "Captain John Chilton to Joseph Blackwell, September 13, 1776," *Tyler's Quarterly Historical and Genealogical Magazine*, Vol. 12, (Richmond, VA: Richmond Press, Inc. 1931), 91.

however, great joy was expressed at our arrival and great things expected from the Virginians, and of consequence we must go through great fatigue and danger.[8]

Chilton noted that they were camped on the northern end of Manhattan Island, near Kings Bridge, about fourteen miles above the town of New York. He worried that since they were on an island, with a massive British fleet offshore, and a large army poised to strike at any moment, they might soon find themselves trapped. He could see the British encampments on Long Island and the flash from their cannon as they fired across the East River. The British maintained a steady bombardment all day, prompting Chilton to complain that he had to write hurriedly.[9]

In the evening a British party was discovered landing on a small island in the East River. As a result, much of the American army, including the 3rd Virginia, manned their alarm posts all night. Captain Chilton and his fellow Virginians marched and countermarched into the early morning hours of September 14th, returning to camp at sunrise. The remainder of the day was relatively quiet, but at midnight they were once again called out to their alarm posts to repeat the actions of the night before.[10]

This time, the alarm was not a false one. On the morning of September 15th, British troops came ashore at Kip's Bay, a mile north of New York City (which occupied just the southern

---

[8] Ibid.
[9] Ibid.
[10] Tyler, ed., "John Chilton to friends, September 17, 1776," *Tyler's Quarterly Historical and Genealogical Magazine*, Vol. 12, 92.

tip of Manhattan Island at the time) and approximately ten miles from the position of the 3rd Virginia. The American troops posted at Kip's Bay offered little resistance and Washington abandoned the city. His troops fled to the high ground of Harlem Heights, near the northern end of the island.

The 3rd Virginia had actually returned to camp on the morning of the 15th, thoroughly exhausted, and was finishing breakfast when word reached them that the enemy had landed on Manhattan Island. It was reported throughout the camp that the American forces posted at Kip's Bay had shamefully abandoned their position without firing a shot.[11] As a result, the town of New York easily fell to the enemy.

The 3rd Virginia once more manned its alarm post in anticipation of an enemy assault. Captain Gustuvus Wallace, commanding a company of 3rd Virginians from King George County, wrote home a few days later saying that,

> *Though our regiment had been out of camp under arms for two nights before, we were ordered to cover the retreat of the cowardly Yankeymen...*[12]

His fellow captain, John Chilton, reported that the men were ready for battle.

> *Our soldiers were greatly exasperated and being drawn up for Battle, it was very discoverable that they*

---

[11] Ibid.
[12] Tyler, ed., "Gustuvus Wallace to his brother, September 17, 1776," *Tyler's Quarterly Historical and Genealogical Magazine*, Vol. 12, 94.

were determined to fight to the last for their country; every soldier encouraging and animating his fellow.[13]

Battle of Harlem Heights

Although the British attack at Kip's Bay caught the Americans by surprise and met little resistance, General Howe once again failed to press his advantage. As a result, most of the American forces stationed in New York City were able to retire northward, to Harlem Heights, before the British severed the roads. Captain Chilton's company, along with the rest of the 3rd Virginia, manned some of the fortifications on Harlem Heights and watched American soldiers straggle in all night. By midnight British troops were posted opposite the Americans at Harlem Heights.

Early the next morning, the British fired upon a detachment of New England rangers who had moved into no-man's land between the two lines.[14] The rangers, commanded by Colonel Thomas Knowlton of Connecticut, held their ground for thirty minutes before retiring in the face of superior numbers.[15] The firing attracted General Washington's attention and he soon arrived to take personal command. Observing an opportunity to exploit the enemy's aggressiveness, he issued orders that placed the Americans on the offensive.

A detachment was ordered forward to re-engage and hold the enemy while a second detachment worked its way around

---

[13] Tyler, ed., "John Chilton to friends, September 17, 1776," *Tyler's Quarterly Historical and Genealogical Magazine,* Vol. 12, 92.

[14] Philip Katcher, "They Behaved Like Soldiers: The Third Virginia Regiment at Harlem Heights", *Virginia Cavalcade,* (Vol. 26, No. 2, Autumn 1976), 64.

[15] Ibid.

the flank, using the woods as cover, to strike the enemy in the rear. If all went as planned, the British advance troops would be cut off from their main line and trapped between two American units.

The 3rd Virginia played an important role in both detachments. Its three rifle companies were attached to the flanking force while the remaining line companies, seven in all, joined the holding force. Captain Chilton, with one of the line companies, described what happened in a letter the next day.

> [In the] *morning we marched down toward* [the enemy] *and posted ourselves near a meadow having that to our front,* [the Hudson] *river to our right, a body of woods in our rear and on our left. We discovered the enemy peeping from their Heights over their fencings and rocks and running backwards and forwards. We did not alter our positions. I believe they expected we should have ascended the hill to them, but finding us still, they imputed it to fear and came down skipping towards us in small parties. At a distance of 250 or 300 yards they began their fire. Our orders were not to fire till they came near but a young officer (of whom we have too many) on the right fired and it was taken up from right to left. We made about 4 fires ... We then all wiped and loaded and sat down in our ranks and let the enemy fire on us near an hour.* [16]

While British attention focused on the Americans in their front, the three Virginia rifle companies, commanded by Major

---

[16] Tyler, ed., "John Chilton to friends, September 17, 1776," *Tyler's Quarterly Historical and Genealogical Magazine*, Vol. 12, 93.

Andrew Leitch – along with Connecticut rangers under Colonel Knowlton -- moved around the enemy to hit them in the rear. Unfortunately, when this force initiated their attack, they were not behind, but rather on the right flank of the British. Despite this miscue, the riflemen and rangers -- joined by the detachment in front of the enemy -- fought with such tenacity that the British were compelled to retreat. They reformed on top of a hill further back. General George Clinton of New York commented that,

> *We* [the attacking Americans] *pursued them to a buckwheat field on top of a high hill, the distance about four hundred paces, where they received a considerable re-enforcement, with several field-pieces, and there made a stand. A very brisk action ensued at this place which continued about two hours. Our people at length worsted them a third time.*[17]

David Griffith, the regimental surgeon for the 3rd Virginia, proudly described the role of the Virginians in a letter home.

> *A very smart action ensued in the true Bush-fighting way....our Troops behaved in a manner that does them the highest Honor. After keeping a very heavy fire on both sides for near three hours they drove the enemy to their main Body and then were prudently ordered to retreat for fear of being drawn into an ambuscade. The 3rd Virga. Regt. was ordered out at the Beginning to maintain a particular post in*

---

[17] Henry P. Johnston, *The Battle of Harlem Heights,* (London: Macmillian, 1897), 80.

front, and Major Leitch was detached with the 3 Rifle Companies to flank the Enemy. He conducted himself on this occasion in a manner that does him the greatest honor and so did his Party, till he received two balls in his Belly and one In his hip...We had 3 men killed and ten wounded. The Loudon [County] *Company suffered most – the Captain behaved nobly. Our whole loss is not yet ascertained. The wounded are not more than 40. Coll. Noleton* [Knowlton] *of the N.E. Rangers is the only officer killed. ...Our Battalion* [after the Riflemen were detached] *were attacked in open field which they drove off and forced them down a Hill...I must mention that the two Yankee Regts. who ran on Sunday fought tollerably well on Monday and in some measure retrieved their reputation. This affair, tho' not great in itself, is of consequence as it gives spirits to the army, which they wanted. Indeed the confusion was such on Sunday that everybody looked dispirited. At present everything wears a different* face.[18]

Captain Chilton was also proud of his men, writing to his brother that,

Our men observed the best order, not quitting their ranks tho exposed to a constant warm fire. I can't say enough in their favour, they behaved like soldiers who fought from principle alone... Tell the old Planters in Fauqr [Fauquier County] *their sons are fine fellows and soldiers.*[19]

---

[18] Johnston, "David Griffith to Major Leven Powell, September 18, 1776," *The Battle of Harlem Heights,* 171-172.

[19] Tyler, ed., "John Chilton to friends, September 17, 1776," *Tyler's Quarterly Historical and Genealogical Magazine*, Vol. 12, 93.

Even General Washington expressed satisfaction with the conduct of the troops, thanking them in the next day's general orders.[20] After defeats at Long Island and Kips Bay, the victorious skirmish at Harlem Heights helped bolster American morale. Although American spirits were dampened by the loss of approximately 100 men killed and wounded and the deaths of Colonel Knowlton and Major Leitch, the fact that they inflicted an equal number of casualties and took the field from the British had a very positive effect on the troops.[21]

The situation on Manhattan between the two armies remained tense, yet no further fighting of significance occurred in September. The 3rd Virginia was briefly attached to General Thomas Mifflin's brigade of Pennsylvanians, but when the 1st Virginia Regiment arrived in late September, General Washington briefly formed a new Virginia brigade under Colonel Weedon's command with the two regiments.[22] The brigade reported nearly 900 officers and men fit for duty.[23] Another 260 men were present but sick and unfit for duty with over fifty sick and absent from camp.[24]

While the 1st and 3rd Virginia Regiments adjusted to their new situation in New York, the 4th, 5th, and 6th Virginia

---

[20] Chase and Grizzard Jr., eds. "General Orders of the Continental Army, September 17, 1776," *The Papers of George Washington, Revolutionary War Series*, Vol. 6, 321.

[21] Johnston, 87.

[22] Chase and Grizzard, Jr., eds. "General Orders September 16, 1776, and October 5, 1776," *The Papers of George Washington, Revolutionary War Series,* Vol. 6, 404-405, 470.

[23] Charles H. Lesser, ed., "General Return of the Army of the United States…September 28, 1776," *The Sinews of Independence: Monthly Strength Reports of the Continental Army*, (Chicago: University of Chicago Press, 1976), 33.

[24] Ibid.

Regiments headed north to join them. On September 3, the Continental Congress ordered three more continental regiments from Virginia northward, specifying that Colonel Adam Stephen's 4th Virginia Regiment be one of them.[25] Congress selected Stephen's unit because he was to be promoted to brigadier-general, (Virginia's third).[26] His experience in the French and Indian War under Washington likely propelled him ahead of Colonel Woodford in the eyes of Congress, much to the chagrin of Woodford.

Arrangements to transport the three regiments by ship up Chesapeake Bay were made and they departed in late September without incident.[27] On September 30, the Continental Congress ordered the regiments to halt in Wilmington, Delaware.[28] Why they were not sent further north to New York is unclear, but they remained in the vicinity of Wilmington and Chester, Pennsylvania until late October when Congress ordered them to march to Trenton, New Jersey.[29]

With five of Virginia's nine continental regiments in the mid-Atlantic states and part of another (the 8th Virginia) in Georgia (and the remainder marching home to recuperate), just three continental regiments remained in the Old Dominion.

---

[25] Ford, "Proceedings of the Continental Congress, September 3, 1776," *Journals of the Continental Congress*, Vol. 5, 733.

[26] Ford, "Proceedings of the Continental Congress, September 4, 1776," *Journals of the Continental Congress*, Vol. 5, 734.

[27] William Morgan, ed., "Virginia Navy Board to General Adam Stephen, September 11, 1776," *Naval Documents of the American Revolution*, Vol. 6, (Washington: Department of the Navy, 1972), 784.

[28] Ford, "Proceedings of the Continental Congress, September 30, 1776," *Journals of the Continental Congress*, Vol. 5, 835.

[29] Philander P. Chase, ed., "Richard Peters to General Washington, October 24, 1776," *The Papers of George Washington, Revolutionary War Series,* Vol. 7, (Charlottesville: University Press of Virginia, 1997), 26.

Virginia's leaders summoned the 7th Virginia Regiment, now under the command of Colonel William Crawford (due to the resignation of Colonel Daingerfield) from Gloucester County to Williamsburg on September 6.[30] With the 9th Virginia Regiment on the eastern shore and the 2nd Virginia Regiment temporarily dissolved due to the expiration of enlistments among its men, the troops of the 7th Virginia were the only Virginia regulars in position to defend the capital.

The Continental Congress offered some relief in mid-September by authorizing the formation of 88 continental battalions, 15 of which were to come from Virginia.[31] Virginia's legislature voted to re-establish the 2nd Virginia and raise six additional continental regiments to meet its quota of fifteen.[32]

The 2nd Virginia was re-constituted without its original colonel, William Woodford. He resigned in protest over the promotion of Adam Stephen (who Woodford argued should have been subordinate to him in rank). Lieutenant Colonel Alexander Spotswood assumed command of the reconstituted 2nd Virginia Regiment.

Another Virginia continental unit that underwent a leadership change was Colonel Hugh Stephenson's combined Virginia-Maryland Rifle Corps. Formed from the remnants of the Virginia and Maryland continental rifle companies of 1775 and supplemented with several new rifle companies from

---

[30] Chase and Grizzard, Jr., eds. "Colonel William Crawford to General Washington, September 20, 1776," *The Papers of George Washington, Revolutionary War Series,* Vol. 6, 350-351.

[31] Ford, "Proceedings of the Continental Congress, September 16, 1776," *Journals of the Continental Congress,* Vol. 5, 762.

[32] Hening, "October 1776," *Statutes at Large, Being a Collection of all the Laws of Virginia...,* Vol 9, 179.

those two states, Stephenson's rifle corps took shape over the summer and fall of 1776. Sadly, Colonel Stephenson never got a chance to command the unit; he grew ill while recruiting in Virginia and died there. Lieutenant Colonel Moses Rawlings of Maryland assumed command.

The Virginia riflemen of this corps did not march as one unit to New York, but rather by separate companies. The first arrived at headquarters in Manhattan in early October.[33] By late October, a portion of the rifle corps was posted at Fort Lee on the west bank of the Hudson River, across from Fort Washington.[34] The two forts were meant to obstruct British navigation up the river. Fort Washington was also the strongest post left on Manhattan Island with American troops.

Rawlings's riflemen were sent across the river in early November to reinforce Fort Washington and it was there that the Virginia and Maryland riflemen shined in a desperate effort to hold back the British and their Hessian allies when they attacked in force on November 16.

A month before the Battle of Fort Washington, British movements on the mainland above Manhattan forced General Washington to remove the bulk of his army from the island and reposition it at White Plains, where a heated, but inconclusive

---

[33] Chase and Grizzard, Jr., eds. "General Washington to Samuel Washington, October 5, 1776," *The Papers of George Washington, Revolutionary War Series,* Vol. 6, 486.
Note: There is a degree of speculation here as General Washington received a letter from his brother Samuel sometime in early October. The letter was delivered by Captain Abraham Shepherd who commanded one of the Virginia rifle companies in Rawlings Virginia-Maryland Rifle Corps.
[34] Philander D. Chase, ed., "General Greene to General Washington, October 29, 1776," *The Papers of George Washington, Revolutionary Series,* Vol. 7, 46.

battle was fought. The Virginia continentals with the main army (the 1st and 3rd Virginia Regiments, who were attached to General Stirling's brigade in mid-October) skirmished with the British in the days leading up to the battle of White Plains, but did not participate significantly in that battle.[35]

Although the Virginians were not heavily engaged in the battle of White Plains, the strain of the campaign began to show on them. The clothing and shoes of many Virginians had worn out and with nothing to replace these items with, the men began to suffer. Colonel Weedon described the hardship the Virginians endured to John Page in late October.

> *The suffering of my poor men makes me feel exceedingly, for these five weeks we have been under arms every morning before day, exclusive of the other necessary duties of the army, which has been uncommonly hard, they have obliged to engage it, entirely naked, some without shoes or stockings, several without blankets and almost all without shirts.*[36]

A troop return for the army in early November reported just 295 officers and men present and fit for duty from the 1st Virginia and 368 such men from the 3rd Virginia. The two regiments had 368 men absent due to illness and another 56 present but unfit due to illness.[37]

---

[35] Chase and Grizzard, Jr., eds. "General Orders, October 17, 1776," *The Papers of George Washington, Revolutionary Series*, Vol. 6, 582-583.

[36] Harry M. Ward, "George Weedon to John Page, October 26, 1776," *Duty, Honor, or Country: General George Weedon and the American Revolution*, (Philadelphia: American Philosophical Society,1979), 66.

[37] Lesser, ed., "General Return of the Army of the United States… November 3, 1776," *The Sinews of Independence: Monthly Strength Reports of the Continental Army*, 37.

The Virginians of the 4th, 5th and 6th Regiments were not included in the troop return as they were still in New Jersey. On November 8, General Stephen informed the Board of War that he was on the march to Amboy from Trenton to join General Mercer. He left 324 convalescents from the three regiments behind but suggested that at least some of them, *"could take a brush with the Enemy occasionally"* if necessary.[38]

While General Stephen's Virginians marched east to Amboy, the 1st and 3rd Virginia Regiments with General Stirling crossed to the west bank of the Hudson River at Haverstraw and prepared to resist any British crossing.[39]

General Howe did indeed intend to cross the river and strike into New Jersey. Before he did so, however, he decided to eliminate the last American post on Manhattan, Fort Washington.

Fall of Fort Washington

Fort Washington, built on a 230 foot elevation overlooking the Hudson River, was originally constructed to challenge British shipping on the river. It was one component of the American defenses at Harlem Heights. When General Washington moved the bulk of his army to White Plains in mid-October, Fort Washington became the bastion of the reduced American presence on Manhattan. Despite General Washington's doubts about the value and security of Fort Washington, he accepted the advice of General Nathanael

---

[38] Adam Stephen, "General Adam Stephen to Board of War, November 8, 1776," *Papers of the Continental Congress*, Vol. 1, 221.
[39] Chase, ed., "Lord Stirling to General Washington November 10, 1776," *The Papers of George Washington, Revolutionary War Series*, Vol. 7, 136-137.

Greene and maintained a garrison there. General Greene, who was posted across the river at Fort Lee, explained the importance of Fort Washington in a letter to General Washington:

> *Upon the whole I cannot help thinking the Garrison* [Fort Washington] *is of advantage – and I cannot conceive the Garrison to be in any great danger the men can be brought off at any time…Col. Magaw* [Fort Washington's commander] *thinks it will take* [the enemy] *till December expires, before they can carry it* [the fort]*…If the Enemy don't find it an Object of importance they wont trouble themselves about it -- if they do, it is full proof they feel an injury from our possessing it – Our giving it up will open a free communication with the Country by the Way of Kings bridge – that must be a great Advantage to them and injury to us.*[40]

Colonel Robert Magaw commanded the 1,200 man garrison at Fort Washington. Although this contingent was adequate to defend the fort, it was far short of the manpower necessary to properly defend the approaches. When Washington realized that Fort Washington was Howe's next objective, he rushed reinforcements there. By mid-November, 2,800 Americans were stationed on Manhattan Island.[41]

---

[40] Chase, ed., "General Nathaniel Greene to General Washington, November 9, 1776," *The Papers of George Washington, Revolutionary War Series,* Vol. 7, 120.

[41] Henry P. Johnson, *The Campaign of 1776 around New York and Brooklyn,* (Brooklyn: Long Island Historical Society, 1878), 277.

Included in that number were approximately 250 riflemen from Virginia and Maryland.[42] They belonged to Colonel Moses Rawlings rifle regiment. Most of the riflemen were new recruits, raised over the summer.

The riflemen were posted along the northern approach to Fort Washington. Virginian Henry Bedinger, a veteran of Captain Hugh Stephenson's rifle company of 1775 that marched to Boston, noted that,

Our Reg't tho weak was most advantageously placed by Rawlings and Williams [Otho] *on a Small Ridge, about half a mile above Fort Washington.*[43]

The riflemen were joined by musket-men from Maryland and Pennsylvania. Other American troops were stationed in the lines at Harlem Heights and along the Harlem River to resist possible attacks from those directions. Colonel Magaw remained at Fort Washington, which was the fall back position for the troops outside the fort.

Despite American efforts to bolster the defenses around Fort Washington, General William Howe was confident that his three pronged, 8,000 man attack, would overwhelm the rebels. Hoping to avoid bloodshed, Howe summoned the Americans to surrender the fort in mid-November. Colonel Magaw rejected the demand and pledged to defend Fort Washington to the last.[44] The stage was set for one of America's worst defeats of the war.

---

[42] Mark M. Boatner, *Encyclopedia of the American Revolution*, (New York: D. McKay Co., 1966), 386.

[43] Dandridge, 157.

[44] George Scheer and Hugh Rankin, eds., *Rebels and Redcoats: The American Revolution Through the Eyes of Those Who Fought and Lived It*, (Da Capo Press, 1957), 198.

The British attack began early in the morning of November 16th. Thousands of British and Hessian troops advanced on Fort Washington from three directions. Two thousand men under General Hugh Percy attacked the American lines at Harlem Heights and easily drove the rebels back. Another three thousand British soldiers crossed the Harlem River under General Charles Cornwallis and swept the American militia aside. Both columns converged on Fort Washington.

The situation north of the fort was different. Colonel Rawlings' riflemen and the supporting musket-men waged a spirited defense against 3,000 Hessians. Henry Bedinger recalled,

> *A few of our men were killed with cannon and grape shot. Not a shot was fired on our side until the enemy had nearly gained the summit. Though at least five times our number, our rifles brought down so many of them that they gave way several times...This obstinacy continued for nearly an hour.*[45]

Hessian Captain Andreas Wiederhold acknowledged the difficulty of the assault:

> *We stood facing their crack troops and their riflemen all on this most inaccessible rock which lay before us, surrounded by swamps and three earthworks, one above the other.*[46]

---

[45] Dandridge, 157.
[46] Scheer & Rankin, 199.

John Reuber, another Hessian soldier, also recalled the difficult fight:

> [We] *marched forward up the hill and were obliged to creep along up the rocks, one falling down alive, another being shot dead. We were obliged to drag ourselves by the beech-tree bushes up the height where we could not really stand.*[47]

The strength of the Hessian attack eventually forced the riflemen back. Henry Bedinger recalled that,

> *Our troops retreated gradually from redoubt to redoubt, contesting every inch of ground, still making dreadful havoc in the ranks of the enemy. We labored, too, under disadvantages, as the wind blew the smoke full in our faces.*[48]

When the riflemen returned to the fort, the flaw in the American defense became apparent. Hundreds of men milled about, unable to find a position along the earthworks. There were too many men in the fort, and many of them had no shelter against bombardment. A Hessian officer summoned Colonel Magaw to spare his men and surrender. Magaw wanted to hold out until evening and attempt a night withdrawal, but his situation was desperate and he capitulated.

The fall of Fort Washington, with over 2,800 men and a significant amount of supplies, was a major blow to the

---

[47] Henry Steele Commager and Richard B. Morris, eds., *The Spirit of 'Seventy-Six: The Story of the American Revolution as Told by Participants*, (Edison, NJ: Castle Books, 2002), 494.
[48] Dandridge, 157.

American army. General Washington acknowledged this in a letter to Congress the day the fort fell:

> *The Loss of such a Number of Officers and Men, many of whom have been trained with more than common Attention, will I fear be severely felt. But when that of the Arms and Accoutrements is added much more so.*[49]

General Howe gave Washington little time to absorb the loss, sending troops across the Hudson River to attack Fort Lee on November 20. General Washington had already begun to evacuate Fort Lee upon the loss of Fort Washington and ordered the removal of military stores to the interior of New Jersey, but a shortage of wagons slowed the process.[50] Washington had also previously ordered General Stirling's brigade (which included the 1st and 3rd Virginia Regiments) to New Brunswick, prior to the loss of Fort Washington.[51] This meant that when Fort Lee fell to the British on November 20, the Virginia continentals in New Jersey were at New Brunswick under General Stirling (the 1st and 3rd Virginia), and at Amboy under General Stephen (the 4th, 5th, and 6th Virginia). A small number of Virginia riflemen joined the retreat from Fort Lee.

---

[49] Chase, ed., "General Washington to the Board of War, November 16, 1776," *The Papers of George Washington, Revolutionary War Series,* Vol. 7, 165.

[50] Chase, ed., "General Washington to John Hancock, November 19-21, 1776," *The Papers of George Washington, Revolutionary War Series,* Vol. 7, 180-182.

[51] Chase, ed., "General Washington to John Hancock, November 14, 1776," *The Papers of George Washington, Revolutionary War Series,* Vol. 7, 154 and Tyler, ed., "John Chilton to his brother Charles Chilton, November 30, 1776," *Tyler's Quarterly Historical and Genealogical Magazine,* Vol. 12, 98.

They were the remnants of Colonel Rawlings's Maryland and Virginia Rifle Corps, about a hundred strong, who were not at Fort Washington when it fell.[52]

General Stirling's division was ordered north to Elizabethtown to meet General Washington's retreating troops while Stephen's brigade remained at Amboy to challenge any potential British landing from Staten Island.[53]

With most of the New England troops (7,500) still at White Plains under General Lee and another 4,000 troops posted in the New York Highlands under General Heath to oppose a possible British advance up the Hudson River, General Washington, with less than 4,000 men, was powerless to stop General Howe's advance into New Jersey.[54] He led the few troops from New Jersey, Pennsylvania, and Maryland that were with him at Fort Lee to Newark and then, after a respite of several days due to poor weather, on to Elizabethtown, where they were met by General Stirling's division on November 27.[55]

With the British pressing them further, General Washington marched on to New Brunswick. The 3rd Virginia Regiment served as part of the rear guard for the retreating Americans. Captain John Chilton described the march to New Brunswick to his brother in a letter dated November 30.

---

[52] Lesser, ed., "General Return of the Army of the United States… December 22, 1776," *The Sinews of Independence: Monthly Strength Reports of the Continental Army*, 43.

[53] Tyler, ed., "John Chilton to his brother Charles Chilton, November 30, 1776," *Tyler's Quarterly Historical and Genealogical Magazine*, Vol. 12, 98.

[54] Lesser, ed., "Return of the Army…December 1, 1776," *The Sinews of Independence: Monthly Strength Reports of the Continental Army*, 40-41.

[55] Tyler, ed., "John Chilton to his brother Charles Chilton, November 30, 1776," *Tyler's Quarterly Historical and Genealogical Magazine*, Vol. 12, 98.

This was a melancholy day, deep miry road and so many men to tread it made it very disagreeable marching, we came 8 or 10 miles and encamped. Yesterday [November 29] we reached this place [New Brunswick]. How long we shall stay, I can't say, but expect we shall make a stand near this placed if not at it, but no certainty when the Enemy are advancing on and an engagement may happen before tomorrow night. We must fight to a disadvantage. They exceed us in numbers greatly.[56]

Aware that his brother would likely be surprised by the significant reversal of American fortunes, Chilton added,

You will wonder what has become of the good army of Americans you were told we had. I really can't tell, they were in some degree imaginary. Militia, some enlisted for 2 some 4, some 5 months, their times were mostly out before the battle of White Plains...and I suspect that the thinness of our Troops was one reason we were not allowed to fight them that day. The same reason prevents us now....[57]

Before Chilton finished his letter, word arrived that the British were moving closer to attack. He speculated that, "*Upon certain intelligence we shall move up I suppose,*" and then declared,

---

[56] Ibid.
[57] Ibid.

Oh god that our Congress should raise men just for an expense till time comes for them to fight and then their time be out! [General] *Howe must have known of this, there are so many Tories all over the Continent. The very time of his landing first was about the time of whole Regmts time being up. Genl. Lee is yet* in [New York] *with 10 or 12,000 but fear he can't join us in time, and indeed, I don't know whether he should come to our assistance. If he should and we get them a little further in the Country we could shortly give a good account of ourselves...I trust, but if the Militia joins us in a day or two, I hope they* [the British] *will repent their bold step. Our men are very willing to fight them on any terms....*[58]

Alas, no reinforcement arrived and even worse, nearly half of General Washington's troops were free to leave the army the next day, December 1, with the completion of their enlistments. Washington appealed to them to stay on, but to no avail.[59] He had no choice but to continue his retreat to Trenton, where he planned to cross the Delaware River and hopefully meet reinforcements.[60]

General Charles Lee commanded some of the reinforcements in New York that General Washington so desperately needed, but he was slow to respond, presenting numerous excuses for his delay.

---

[58] Ibid.
[59] Chase, ed., "General Washington to John Hancock, December 1, 1776," *The Papers of George Washington, Revolutionary War Series,* Vol. 7 244.
[60] Chase, ed., "General Washington to John Hancock, December 1, 1776, ½ after 7 p.m." *The Papers of George Washington, Revolutionary War Series,* Vol. 7, 245.

Future Chief Justice John Marshall described the situation faced by General Washington at the start of December in his biography of Washington years later.

> *General Washington found himself at the head of* [a] *small band of soldiers, dispirited by their losses and fatigues, retreating almost naked and barefooted, in the cold of November and December, before a numerous, well appointed and victorious* [enemy].[61]

With General Washington in early December 1776, were five Virginia continental regiments, serving as a rear guard for the retreating army. They halted in Princeton, twelve miles from Trenton, while the American baggage and military stores were carried across the river.[62] With the advance of the British on December 7, General Washington withdrew the rear guard to Trenton and then across the Delaware River.[63] With most of the boats in the area collected and secured on the west bank of the river, it appeared the Americans were momentarily safe.

---

[61] John Marshall, *The Life of George Washington*, Vol. 2, (Fredericksburg, VA: The citizens Guild of Washington's Boyhood Home, 1926), 234.

[62] Chase, ed., "General Washington to John Hancock, December 5, 1776," *The Papers of George Washington, Revolutionary War Series*, Vol. 7, 262.

[63] Chase, ed., "General Washington to John Hancock, December 8, 1776," *The Papers of George Washington, Revolutionary War Series*, Vol. 7, 273.

New York & New Jersey

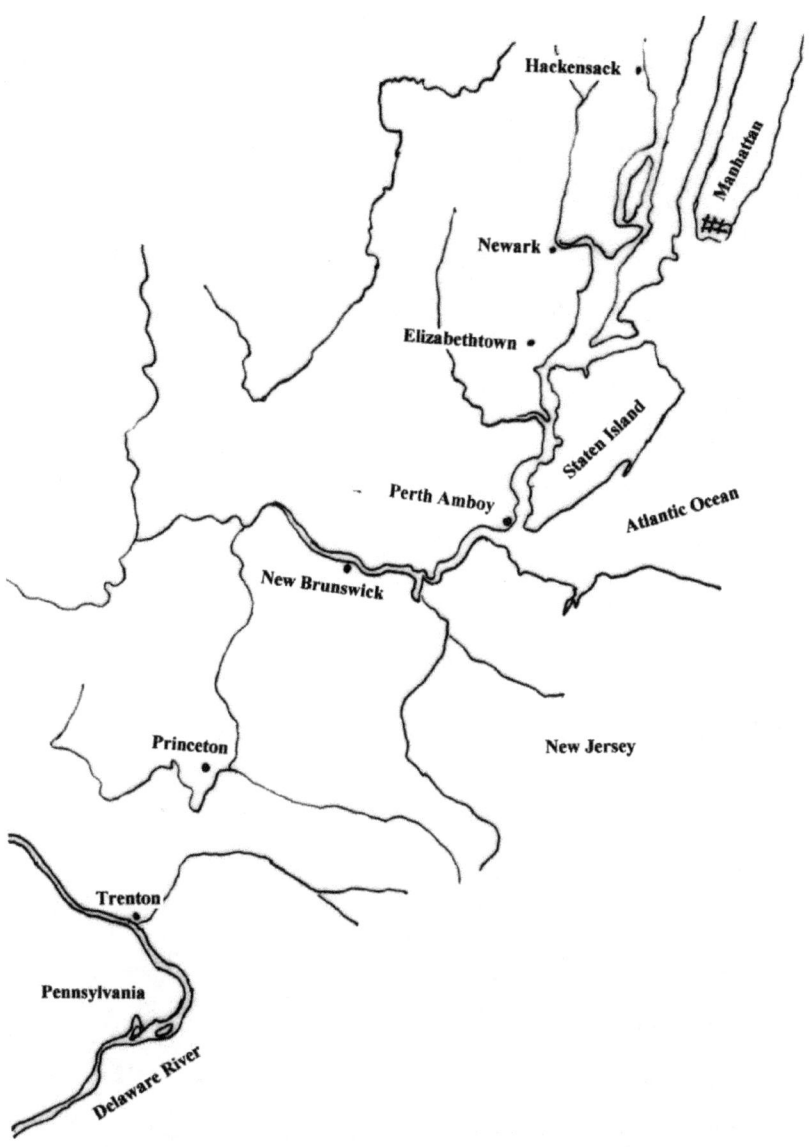

Chapter Six

Trenton and Princeton

December 1776 -- January 1777

Lacking proper shelter and winter clothing, Washington's men stood watch along the banks of the Delaware River during the frigid days and nights of December. Lieutenant Enoch Anderson of the Delaware Regiment (which was attached to General Stirling's brigade) recalled that

> *We lay amongst the leaves without tents or blankets, laying down with our feet to the fire. It was very cold. We had meat, but no bread. We had nothing to cook with but our ramrods, which we run through a piece of meat and roasted it over the fire, and to hungry soldiers it tasted sweet.*[1]

Concerned that General Howe might find a way to cross the river above Trenton, General Washington posted General Stirling's, Stephen's, Mercer's, and de Fermoy's brigades upriver, along a five mile stretch of the Delaware between Coryell's Ferry [New Hope, PA] and McKonkey's Ferry [Washington Ferry, PA].[2] *"Everything in a manner depends on a defence at the Waters Edge,"* stressed Washington, so, *"one*

---

[1] Henry Hobart Bellas, ed., *Personal Recollections of Captain Enoch Anderson, an Officer of the Delaware Regiments in the Revolutionary War*, (Wilmington: The Historical Society of Delaware, 1896), 28.
[2] Chase, ed., "General Orders December 12, 1776," *The Papers of George Washington, Revolutionary War Series,* Vol. 7, 303.

brigade is to support another without loss of time, or waiting orders from me".[3]

General Stirling's brigade was posted on the left end of the four brigades and Stirling reported to General Washington on December 12 that,

> *I have sent one piece of Cannon to Colonel Wieden* [Weedon] *and as the three Regiments* [1st and 3rd Virginia and the Delaware Regiment] *here now lie Compact & well Covered with Boards and Nearly Centrical to Yardly's and Corriels ferries I believe it be best to let them be in their present situation 'till some Movement of the Enemy makes it Necessary to Alter it.*[4]

Although it appears from General Stirling's letter that his troops were adequately sheltered from the weather, Captain Chilton of the 3rd Virginia recalled that the lack of clothing and a common ailment among the troops made for a very difficult time for the men.

> *The weather was extremely cold and duty hard, when we encamped at Blue Mount* [Bowman's Hill] *the men bare of clothes and to a man we all had* [the camp itch].[5]

---

[3] Chase, ed., "General Washington to Generals James Ewing, Hugh Mercer, Adam Stephen, and Lord Stirling, December 14, 1776," *The Papers of George Washington, Revolutionary War Series,* Vol. 7, 331-332.

[4] Chase, ed., "General Stirling to General Washington December 12, 1776," *The Papers of George Washington, Revolutionary War Series,* Vol. 7, 319.

[5] Tyler, ed., "Captain John Chilton to his brother Charles Chilton, February 11, 1777," *Tyler's Quarterly Historical and Genealogical Magazine,* Vol. 12, 112.

It is reasonable to expect that the Virginians under Generals Stephen and Mercer (whose brigade included the remnants of Rawling's Rifle Corps) experienced similar hardship.

David Griffith, the regimental surgeon of the 3rd Virginia, did not comment on the condition of the men, but rather, lamented the lack of support given the army by the populace:

> *We have much need for a speedy re-inforcement. I am much afraid we shall not have it in time to prevent the destruction of American affairs... Everything here wears the face of despondency...A strange consternation seems to have seized everybody in this country. A universal dissatisfaction prevails, and everybody is furnished with an excuse for declining the publick service.*[6]

General Washington also noted the universal dissatisfaction displayed by the local populace and exclaimed that if help did not arrive soon the American cause might be lost:

> *A large part of the Jerseys have given every proof of disaffection that a people can do, & this part of Pennsylvania are equally inimical; in short your imagination can scarce extend to a situation more distressing than mine -- Our only dependence now, is upon the Speedy Inlistment of a New Army, if this fails us, I think the game will be pretty well up, as from disaffection and want of spirit & fortitude, the Inhabitants instead of resistance, are offering Submission, & taking protections from Genl Howe in Jersey.*[7]

---

[6] Tyler, ed., "David Griffith to Major Powell, December 8, 1776," *Tyler's Quarterly Historical and Genealogical Magazine*, Vol. 12, 101.
[7] Chase, ed., "General Washington to Lund Washington, December 10-17,

Washington was equally candid about the army's prospects with his brother Samuel:

> *Between you and me I think our Affairs are in a very bad way.... I have no doubt that General Howe will still make an attempt upon Philadelphia this Winter – I see nothing to oppose him in a fortnight from this time, as the term of all the Troops except those of Virginia (reduced to almost nothing) and Smallwood's Regiment from Maryland (in the same condition) will expire in that time. In a word, my dear Sir, if every nerve is not straind to recruit the New Army with all possible Expedition I think the game is pretty near up....*[8]

General Washington was so discouraged at the course of events that he instructed Lund Washington to prepare to flee Mount Vernon:

> *Matters to my view, (but this I say in confidence to you, as a friend) wear so unfavourable an aspect...that I would look forward to unfavourable Events, & prepare Accordingly in such a manner however as to give no alarm or suspicion to any one; as one step towards it, have my Papers in such a Situation as to remove at a short notice in case an Enemy's Fleet should come up*

1776," *The Papers of George Washington, Revolutionary War Series*, Vol. 7, 291.

[8] Chase, ed., "General Washington to Samuel Washington, December 18, 1776," *The Papers of George Washington, Revolutionary War Series*, Vol. 7, 370.

the River – When they are removed let them go immediately to my Brothers in Berkeley.[9]

The stunning news of General Charles Lee's capture by the British on December 14th, near Basking Ridge, New Jersey further dampened American morale. Lee had foolishly strayed a few miles from his troops during their slow march to join Washington and was seized by a party of British dragoons that were tipped off about Lee's location by local Tories.

The loss of General Lee did not stop the 2,000 desperately needed reinforcements that marched with him from finally joining General Washington in New Jersey. Their arrival, and another 600 troops under General Horatio Gates, coupled with reports that General Howe had ceased major operations at least until the Delaware River froze solid, eased Washington's immediate concern and presented him with an opportunity.[10]

To provide adequate shelter for his men for the winter, General Howe dispersed his British and Hessian troops among several towns in New Jersey in mid-December and then returned to New York City. The overconfident British commander posted only 1,400 Hessian troops in Trenton under Colonel Johann Rall. It was this force that caught the attention of Washington and several other American officers.[11]

---

[9] Chase, ed., "General Washington to Lund Washington, December 10-17, 1776," *The Papers of George Washington, Revolutionary War Series*, Vol. 7, 291.

[10] Chase, ed., "General Washington to Robert Morris, December 22, 1776," *The Papers of George Washington, Revolutionary War Series*, Vol. 7, 412.

[11] David Hackett Fischer, *Washington's Crossing*, Appendix H, (Oxford University Press: 2004), 396.

One of those officers was Colonel Joseph Reed, a former aide to Washington in 1775 and a prominent leader in Pennsylvania. He urged Washington in a letter dated December 22, to take bold action, declaring to the American commander that, *"Some Enterprize must be undertaken in our present Circumstances or we must give up the Cause."*[12] Reed noted that more than half of the troops with Washington were due to depart at the end of December and urged Washington to act decisively while he still could, suggesting that, *"Will it not be possible my dear Genl. for your Troops...to make a Diversion or something more at or about Trenton?*[13] Reed concluded his letter by noting that, *"Our Affairs are* [hastening] *fast to Ruin if we do not retrieve them by some happy Event. Delay with us is now equal to a total Defeat."*[14]

General Washington received Reed's letter late in the day on December 22, and held a council of war that evening. Although no record of the war council exists, Washington revealed the outcome of the meeting to Colonel Reed.

> *Christmas day at Night, one hour before day is the time fixed upon for our Attempt on Trenton. For heaven's sake keep this to yourself, as the discovery of it may prove fatal to us, our numbers, sorry I am to say, being less than I had any conception of – but necessity, dire necessity will – nay must justify any* [attempt].[15]

---

[12] Chase, ed., "Colonel Joseph Reed to General Washington, December 22, 1776," *The Papers of George Washington, Revolutionary War Series,* Vol. 7, 415.

[13] Ibid.

[14] Ibid.

[15] Chase, ed., "General Washington to Colonel Joseph Reed, December 23, 1776," *The Papers of George Washington, Revolutionary War Series,* Vol. 7, 423.

The attempt that General Washington had in mind was against the 1,400 Hessians posted in Trenton. Washington planned to send nearly every solider available, some 5,000 troops, across the Delaware River in three simultaneous night crossings along a twenty-mile front.

General John Cadwalader, with approximately 1,000 Pennsylvania militia and 850 New England continentals that Lee and Gates had recently brought to the army, was ordered to cross the river ten miles south of Trenton to engage the enemy around Burlington and prevent their relief of Trenton.[16] General James Ewing, with approximately 1,100 militia from Pennsylvania and New Jersey was ordered to cross the river at Trenton Ferry (just south of Trenton) to prevent the Hessian garrison at Trenton from fleeing southward once General Washington attacked the town from the north.[17]

Washington himself planned to cross the river about nine miles above Trenton with the troops he had led into Pennsylvania earlier in the month and the recently arrived troops of General Sullivan (formerly Lee). Leaving the sick and infirm as well as a baggage guard behind in camp, the force directly under General Washington's command numbered approximately 2,400 men. They were divided into two divisions under General Sullivan and General Greene. The Virginia continental regiments in the brigades of Generals Stirling and Stephen, and Virginia riflemen, which had been attached to General Mercer's brigade, were all assigned to General Greene's division.[18]

---

[16] William S. Stryker, *The Battles of Trenton and* Princeton, (Old Barracks Association, 2001), 344-46.
[17] Ibid. 346-47.
[18] Chase, ed., "General Orders, December 25, 1776," *The Papers of George Washington, Revolutionary War Series,* Vol. 7, 434-436.

A troop return completed just a few days before the attack revealed that the two Virginia regiments under General Stirling numbered 366 officers and men fit for duty, (185 from the 1st Virginia and 181 from the 3rd Virginia).[19] General Stephen's brigade of Virginians totaled 549 officers and men fit for duty (229 from the 4th Virginia, 129 from the 5th Virginia and 191 from the 6th Virginia).[20] Just over one hundred riflemen were fit for duty in Colonel Rawling's combined Maryland and Virginia rifle corps attached to General Mercer's brigade.[21] It is impossible to distinguish how many of these officers and men were Virginia riflemen.

General Washington issued orders to his brigade commanders on Christmas Eve outlining the attack. They in turn issued more detailed orders to their brigades. General Mercer's orders to Colonel John Durkee of the 20th Continental Regiment reveal the preparations likely made in every regiment for the attack on Trenton.

> *You are to see that your men have three days provisions ready cooked before 12 o' clock this forenoon – the whole fit for duty except a Serjeant and six men to be left with the baggage, and to parade precisely at four in the afternoon with their arms, accoutrements & ammunition in the best order, with their provisions and blankets – you will have them told off in divisions in which order they are to march – eight men a breast, with the officers fixed to their divisions from which they are on no account to separate –*

---

[19] Lesser, ed., "General Return of the Army of the United States… December 22, 1776," *The Sinews of Independence: Monthly Strength Reports of the Continental Army*, 43.
[20] Ibid.
[21] Ibid.

no man is to quit his division on pain of instant punishment – each officer is to provide himself with a piece of white paper stuck in his hat for a field mark. You will order your men to assemble and parade them in the valley immediately over the hill on the back of McConkey's Ferry, to remain there for farther orders – a profound silence is to be observed, both by officers and men, and a strict and ready attention paid to whatever orders may be given....[22]

General Washington selected General Stephen's brigade to serve as the advance guard of the army.[23] They crossed the river, accompanied by General Washington, soon after sunset of Christmas Day and formed a perimeter around the landing area. With the weather deteriorating and ice forming on the river, transporting the troops, cannon, and horses across the Delaware took longer than anticipated. Major James Wilkinson described the effort in his memoirs years later.

The force of the current, the sharpness of the frost, the darkness of the night, the ice...and high wind, rendered the passage of the river extremely difficult.[24]

John Greenwood, a young fifer from Massachusetts in General John Sullivan's Division, recalled that, "*it rained, hailed, snowed and froze, and at the same time blew a perfect*

---

[22] Stryker, "General Mercer to Colonel Durkee, December 25, 1776," *The Battles of Trenton and Princeton*, 379.

[23] Chase, ed., "General Orders, December 25, 1776," *The Papers of George Washington, Revolutionary War Series*, Vol. 7, 434-436.

[24] James Wilkinson, *Memoirs of My Own Times*, Vol. 1, (Philadelphia: Abraham Small, 1816), 128.

hurricane."[25] Like the Virginians in General Greene's division, Greenwood struggled to stay warm around a large bonfire while he waited on the Jersey side of the river for the rest of the army and equipment to cross. He remembered that

> *When I turned my face towards the fire my back would be freezing.... By turning round and round I kept myself from perishing....*[26]

General Washington, impatiently waiting on the east bank of the river for the last of his troops to cross, grew increasingly concerned that the delay would cost them the element of surprise.[27] Washington recalled to Congress that as the night wore on

> [I began to] *Despair of surprising the Town, as I well knew we could not reach* [Trenton] *before the day was fairly broke, but as I was certain there was no making a Retreat without being discovered, and harassed on repassing the River, I determined to push on at all Events.*[28]

The two other American detachments downriver from Trenton involved in the attack faced equally difficult conditions. Stymied by poor weather and growing ice, General Cadwalader and General Ewing suspended their assaults after

---

[25] Isaac J. Greenwood, ed., *The Revolutionary Services of John Greenwood... 1775-1783*, (New York, 1922), 39.
[26] Ibid.
[27] Fischer, *Washington's Crossing*, 219.
[28] Chase, ed., "General Washington to John Hancock, December 27, 1776," *The Papers of George Washington*, Vol. 7, *Revolutionary War Series*, 454.

struggling for hours to cross the river. Both assumed that General Washington had met a similar fate. Washington, unaware of developments to the south, carried on.

The last of Washington's eighteen cannon were unloaded on the Jersey shore sometime after 3 a.m. and the half frozen American troops began a nine-mile march to Trenton in one long column. Fifer Greenwood of General Sullivan's division, recalled that

> *We began an apparently circuitous march, not advancing faster than a child ten years old could walk, and stopping frequently.... During the whole night it alternately hailed, rained, snowed, and blew tremendously...*[29]

General Stephen's Virginians led the American column for the first half of the march. Ensign Robert Beale of the 5th Virginia Regiment recalled in his memoirs that the night, "*became very tempestuous, rain, hail, and snow, but we marched all night.*"[30]

Beale must have been exhausted for he and a detachment of Stephen's men had been across the river in New Jersey hours before General Washington and the army crossed. Ensign Beale recalled that they had been ordered across the river earlier on Christmas Day by General Stephen to "*show ourselves to the Hessian picket,*" which they did, alarming the Hessians

---

[29] Greenwood, ed., *The Revolutionary Services of John Greenwood... 1775-1783*, 39.

[30] Dennis P. Ryan, "Robert Beale Memoirs," *A Salute to Courage: The American Revolution as Seen Through Wartime Writings of Officers of the Continental Army and Navy*, (NY: Columbia University Press, 1979), 56.

guarding Trenton.[31] Beale remembered that, *"We immediately retreated with all speed,"* and the detachment made their way back upriver, where they encountered Washington's army crossing the river.[32]

Historian Harry Ward noted in his biography of General Stephen that Stephen had sent the detachment across the river to exact revenge for the death of one of his soldiers and, claimed Stephen, they succeeded, killing four and wounding several others.[33] If any losses were inflicted on the Hessians, they paled in comparison to the potential loss of surprise caused by Stephen's unauthorized detachment. Ward noted that when General Washington learned what had occurred he became enraged, exclaiming to Stephen, *"You Sir, may have ruined all my plans."*[34] Undeterred, Washington pressed on, ordering his troops forward to Trenton at 4:00 a.m., four hours behind schedule.[35]

About halfway to Trenton, Washington's column split. General Sullivan led his division along the River Road to strike Trenton from the west while General Greene led his division along the Pennington Road to strike Trenton from the north. The orders for each detachment were simple and direct, force the enemy guard and, *"push directly into the Town, that they might charge the Enemy before they had time to form."*[36]

---

[31] Ibid., 55.
[32] Ibid.
[33] Harry M. Ward, *Major General Adam Stephen and the Cause of American Liberty*, (Charlottesville: Univ. Press of Virginia, 1989), 151.
[34] Ibid.
[35] Chase, ed., "General Washington to John Hancock, December 27, 1776," *The Papers of George Washington,* Vol. 7, *Revolutionary War Series,* 454.
[36] Ibid.

Trenton and Princeton

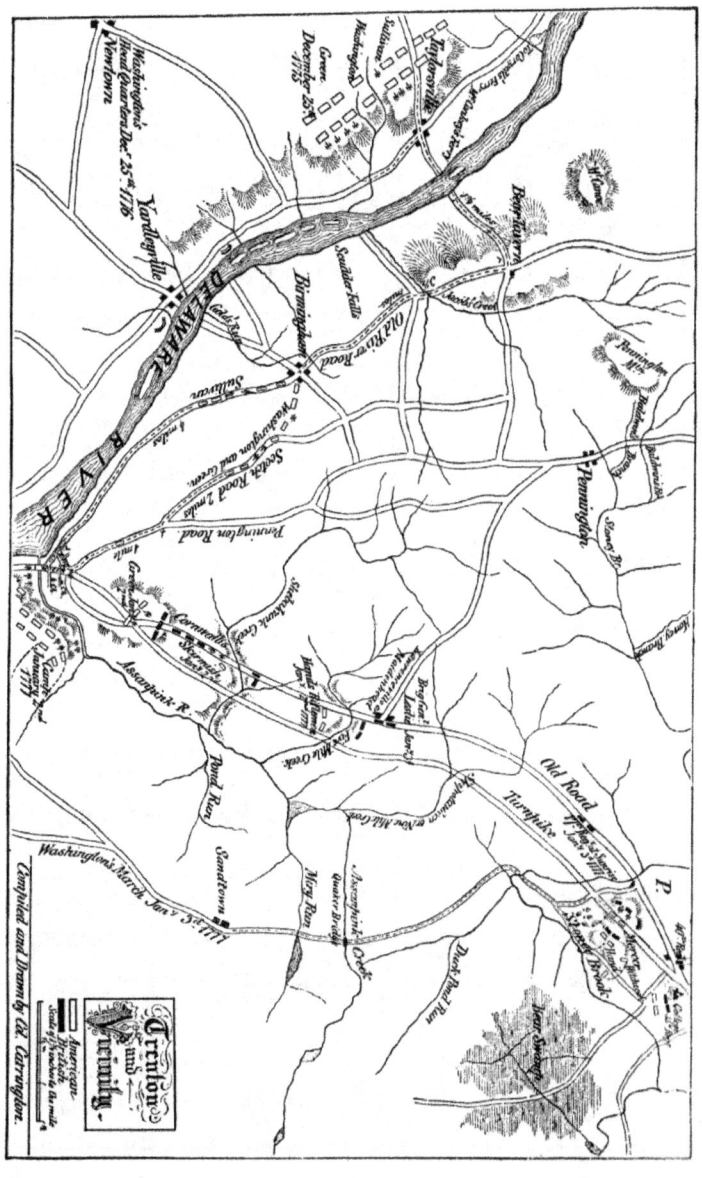

General Washington described what occurred when contact with the enemy pickets occurred.

> *The upper division* [General Greene's] *arrived at the Enemy's advance post, exactly at eight OClock, and in three Minutes after; I found from the fire on the lower Road that, that Division had also got up. The Out Guards made but small Opposition, tho' for their Numbers, they behaved very well, keeping up a constant retreating fire from behind Houses. We presently saw their main Body formed, but from their Motions, they seem'd undetermined how to act.*[37]

Washington's assessment of the Hessians was correct; surprised by the sudden American onslaught from two directions, the Hessian officers and soldiers fell into confusion. Ensign Beale credited, *"the inclemency of the weather* [for throwing] *the enemy entirely off guard."*[38] The Hessians were stunned that the Americans had managed to cross the river and march down upon them undetected in the midst of the storm, but it was the storm that provided Washington's troops with cover (and tormented them during the march).

Colonel Henry Knox commanded the American artillery at Trenton and positioned several pieces at the head of the town's two main streets. He described the effect of his cannon on the Hessians:

---

[37] Ibid.
[38] Ryan, "Robert Beale Memoirs," *A Salute to Courage: The American Revolution as Seen Through Wartime Writings of Officers of the Continental Army and Navy*, 56.

> *The hurry, fright, and confusion of the enemy was (not) unlike that which will be when the last trumpet shall sound. They endeavoured to form in streets, the heads of which we had previously the possession of with cannon and howitzers; these, in a twinkling of an eye, cleared the streets. The backs of the houses were resorted to for shelter. These proved ineffectual:* [our] *musketry soon dislodged them.*[39]

According to John Greenwood, who was with General Sullivan's division on the River Road, it was not American musketry that dislodged the enemy but sheer numbers:

> *As we had been in the storm all night we were not only wet through and through ourselves, but our guns and powder were wet also, so that I do not believe one would go off and I saw none fired by our party.... We advanced, and although there was not more than one bayonet to five men, orders were given to, 'Charge bayonets and rush on!' and rush on we did. Within pistol-shot they again fired point-blank at us; we dodged and they did not hit a man, while before they had time to reload we were within three feet of them, when they broke in an instant and ran like so many frightened devils into town, which was a short distance, we after them pell-mell.*[40]

The Hessian commander, Colonel Johann Rall, rallied his men in an orchard east of town and attempted to attack Washington's left flank near the Princeton Road, but General

---

[39] Ibid.
[40] Greenwood, ed., *The Revolutionary Services of John Greenwood...* 1775-1783, 41.

Washington acted quickly and shifted more troops to that position to secure it.[41]

With large numbers of hostile troops on his front and flanks, Colonel Rall's next move should have been an attempted retreat across the Assunpink Creek. Instead, the proud Hessian commander ordered his men back into town to recapture two cannon.[42] When a party of Hessians approached the abandoned cannon, Captain William Washington of the 3rd Virginia charged forward. Major James Wilkinson described what happened.

> *Captain Washington, who, seconded by Lieutenant James Monroe, led the advanced guard of the left column, perceiving that the enemy were endeavouring to form a battery, rushed forward, drove the artillerists from their guns, and took two pieces in the act of firing.*[43]

Sergeant Joseph White, a New England artillerist who joined Captain Washington's assault with a detachment of artillerymen recalled that

> *I hallowed as loud as I could scream to the men to run for their lives right up to the* [cannon]. *I was the first that reach them.* [The Hessians] *had all left except one man tending the vent.*[44]

---

[41] Fischer, *Washington's Crossing*, 246.
[42] Ibid.
[43] James Wilkinson, *Memoirs of My Own Times*, Vol. 1, (Philadelphia: Abraham Small, 1816), 130.
[44] Fischer, *Washington's Crossing*, 247.

White and his men overwhelmed and captured the remaining Hessian artillerist and turned the captured guns on the enemy. *"We put in a canister of shot (they had put in a cartridge before they left it) and fired,"* recalled Sergeant White.[45] Captain Washington and Lieutenant Monroe were both seriously wounded in the charge, but the Hessian guns were secured and the battle approached its dramatic conclusion.

The loss of their cannon, as well as their commander, Colonel Rall, (who was mortally wounded upon his horse) dispirited the Hessians. They withdrew back to the orchard, closely pursued by the Americans who pressed them on three sides. Trapped by the Assunpink Creek in their rear, and the Americans on their front and flanks, the Hessians had little choice but to surrender. They had suffered over one hundred casualties in the battle while the Americans lost just a handful of men.[46]

General Washington's attack on Trenton was a staggering success, garnering over 900 Hessian prisoners along with much needed supplies.[47] More importantly, the victory provided a huge boost to American morale.

The threat of a British counterattack from Princeton and Bordentown caused Washington to immediately march his weary army back to McKonkey's Ferry to re-cross the river. His exhausted troops literally collapsed in camp upon their return in the evening, but they rested with a strong sense of accomplishment; their victory at Trenton restored hope in the American cause.

---

[45] Ibid.
[46] Stryker, *The Battles of Trenton and* Princeton, 196, 194.
[47] Ibid. 386.

News of Washington's victory spread quickly and revived the flagging American morale. While his troops rested, Washington learned that General Howe had withdrawn his troops to central New Jersey. The American commander decided to fill part of the vacuum left by their departure. On December 30, Washington's army re-crossed the Delaware River and occupied Trenton. News of their crossing prompted General Howe to move with uncharacteristic speed, sending approximately 8,000 men under General Charles Cornwallis to destroy Washington's revived army.[48] The stage was set for round two.

2nd Battle of Trenton

General Washington expected the British to challenge him at Trenton and positioned his men along high ground on the bank of the Assunpink Creek, south of town. He concentrated his artillery on a bridge that spanned the creek. On January 2nd, reinforcements arrived and were placed under General Cadwalader swelling Washington's army to approximately 6,000 men.[49]

Reports that the British were gathering in Princeton prompted General Washington to place a strong detachment along the Princeton Road to harass and delay their advance. General de Fermoy commanded these troops, which included detachments from General Stephen's three Virginia regiments (the 4th, 5th and 6th Virginia Regiments). A battalion of Pennsylvania Germans, Colonel Edward Hand's 1st Continental

---

[48] Samuel S. Smith, *The Battle of Princeton*, (Monmouth Beach, NJ: Philip Freneau Press, 1967), 12.
[49] Ibid.

riflemen, and six field pieces also made up General de Fermoy's blocking force.[50]

They were posted along Shabbakonk Creek, about three miles northeast of Trenton. Advance parties were placed at a small creek called Five Mile Run and in the village of Maidenhead (present day Lawrenceville). When the British began their march toward Trenton at daybreak of January 2nd, these American outposts were the first ones they encountered.

The American pickets at Maidenhead drew first blood when they unhorsed a jaeger officer who had ridden in advance of Cornwallis's vanguard. Realizing that a large enemy force was approaching, the pickets withdrew to Five Mile Run, joined the riflemen stationed there in a brief stand, and then resumed their retreat to Shabbakonk Creek.[51]

The American troops at the creek were in turmoil because their commander, General de Fermoy, had inexplicably abandoned them and returned to Trenton. Colonel Edward Hand assumed command and led the first serious resistance to the British advance. His men were well concealed in a heavily forested position. The British, in contrast, advanced over mostly open terrain. Colonel Hand's men held their fire until the enemy came within point blank range and then unleashed a devastating volley upon the British advance guard. They then maintained their fire while the British recoiled and deployed their main force. This was exactly what the American blocking force was instructed to do, delay the British advance, and Major James Wilkinson noted that,

---

[50] Ryan, "Robert Beale Memoirs," *A Salute to Courage: The American Revolution as Seen Through Wartime Writings of Officers of the Continental Army and Navy*, 56, and Wilkinson, *Memoirs of My Own Times*, Vol. 1, 135.

[51] Wilkinson, *Memoirs of My Own Times*, Vol. 1, 136.

> *This operation,* [at the Shabbakonk] *consumed two hours, during which time the rifle corps took breadth and were ready to renew the attack.*[52]

Colonel Hand's men slowly withdrew, resisting all the way to Trenton, bringing Cornwallis's advance to a crawl. When they passed a ravine on the outskirts of town, they joined General Stephen's detachment of Virginians (under the command of Colonel Charles Scott) in another stand. General Washington arrived on the scene with about two hours of daylight remaining. Major Wilkinson recalled,

> *General Washington…feeling how important it was to retard the march of the enemy until nightfall…thanked the detachment, and particularly the artillery, for the services of the day, gave orders for as obstinate a stand as could be made on that ground, without hazarding the* [artillery] *pieces, and retired to marshal his troops for action, behind the Assanpink."*[53]

A thirty-minute artillery duel delayed the British, but they eventually overwhelmed the Americans and continued their advance, forcing the Americans back to Assunpink Creek. Ensign Robert Beale of the 4th Virginia recalled,

> [Major Forsyth] *ordered to the right about face on and off in order. We had not taken more than regular steps until the word, 'Shift for yourselves, boys, get over the*

---

[52] Ibid.
[53] Ibid, 138.

bridge as quick as you can.' There was running followed by a tremendous fire from the British.[54]

The Americans raced through town and across the Assunpink Bridge, halting on the other side to make one last desperate stand. The gravity of the moment weighed heavy on everyone. Ensign Beale recalled,

> *This was a most awful crisis. No possible chance of crossing the* [Delaware] *river; ice as large as houses floating down, and no retreat to the mountains, the British between us and them. Our brigade, consisting of the Fourth, Fifth, and Sixth Virginia Regiments, was ordered to form in column at the bridge and General Washington came and, in the presence of us all, told Colonel Scott to defend the bridge to the last extremity. Colonel Scott answered with an oath, 'Yes, General, as long as there is a man alive.'*[55]

Major Wilkinson also noted the urgency of the situation:

> *If ever there was a crisis in the affairs of the revolution,"* he recalled, *"this was the moment; thirty minutes would have sufficed to bring the two armies into contact, and thirty more would have decided the combat...*[56]

---

[54] Ryan, "Robert Beale Memoirs," *A Salute to Courage: The American Revolution as Seen Through Wartime Writings of Officers of the Continental Army and Navy*, 56.
[55] Ibid.
[56] Wilkinson, *Memoirs of My Own Times*, Vol. 1, 138.

Three times Hessian and British troops advanced towards the bridge, and each time a barrage of American artillery and small arms fire forced them back with heavy losses. The Virginians would not budge, and with the last rays of daylight fading in the west, General Cornwallis decided to suspend the attack and resume it in the morning. General Washington and his men had earned a twelve-hour reprieve, and they made full use of it.

Around midnight, after a few hours of tense rest, most of the American army quietly withdrew from the lines and marched along a little used back road towards Princeton. Washington hoped to surprise the small British garrison there with a dawn attack. The maneuver required stealth and deception, so the men were ordered to keep silent. About 400 area militia remained in the lines at Trenton to maintain the appearance of an army preparing for battle.[57] They kept the campfires burning and continued to dig earthworks to convince the British that Washington was still there. Major Wilkinson reported that General Washington

> *Ordered the guards to be doubled, a strong fatigue party to be set to work on an intrenchment...within hearing of the sentinels of the enemy, the baggage to be sent to Burlington, the troops to be silently filed off by detachments, and the neighboring fences to be used for fuel to our guards, to keep up blazing until toward day when they had orders to retire. The night, although cloudless, was exceedingly dark, and, though calm, most severely cold, and the movement was so*

---

[57] Stryker, *The Battles of Trenton and* Princeton, 275.

cautiously conducted as to elude the vigilence of the enemy.[58]

For the most part the ruse worked. Only a handful of British sentries reported movement in the American camp, but these reports went unheeded.

The route the Americans took to Princeton barely qualified as a road and was very difficult on the horses and men. One soldier recalled,

> *The horses attached to our cannon were without shoes, and when passing over the ice they would slide in every direction and could advance only by the assistance of the soldiers. Our men too, were without shoes or other comfortable clothing; and as traces of our march towards Princeton, the ground was literally marked with the blood of the soldiers' feet.*[59]

Battle of Princeton

As the Americans approached the outskirts of Princeton, General Washington split his force. General Greene was sent to the left to secure a bridge at Stony Brook and enter Princeton along the Post Road, while General Sullivan continued along the back road with the bulk of the army. General Greene's smaller column consisted of approximately 350 troops under General Hugh Mercer, which included the remnants of his brigade as well as the few Virginia continentals from the 1st

---

[58] Wilkinson, *Memoirs of My Own Times*, Vol. 1, 140.
[59] Sergeant R, "The Battle of Princeton," *The Pennsylvania Magazine of History and Biography*, Vol. 20, No. 1 (1896), 515.

Virginia Regiment who were still fit for duty.[60] A much larger brigade under General John Cadawalder with approximately 1,150 men followed Mercer's troops.[61]

General Stephen's Virginians marched in Sullivan's column and the 3rd Virginia regiment had yet to return from escorting the Hessian prisoners of Trenton to Philadelphia.[62]

At almost the same moment that Washington divided his army, a British column about a mile to the west crossed the Stony Brook Bridge and ascended a hill on their way to Trenton. They were reinforcements (over 400) from the 17th and 55th British Regiments under Lieutenant Colonel Charles Mawhood.[63] As they climbed the hill, some of Mawhood's horsemen caught a glimpse of Sullivan's column moving towards Princeton. Mawhood could not determine the size of the American force, but realized that the lone British regiment left in Princeton, the 40th, was in danger, so he reversed direction and rapidly marched back to town. General Washington, who was with Sullivan, soon learned about Mawhood's column. He assumed that it was only a British reconnaissance force from Princeton and ordered General Mercer to pursue and attack it before it warned the town's

---

[60] Fischer, *Washington's Crossing,* 408.
[61] Wilkinson, *Memoirs of My Own Times*, Vol. 1, 141. See also: Caesar Rodney, ed., *The Diary of Captain Thomas Rodney, 1776-1777,* (Wilmington: The Historical Society of Delaware, 1888), 33, and Fischer, *Washington's Crossing,* 408, and Samuel Smith, *Battle of Princeton,* 34.
[62] Ward, *Duty, Honor, or Country: General George Weedon and the American Revolution,* 75-76.
[63] Smith, *Battle of Princeton,* 19.

garrison.[64] Mercer responded quickly; he marched his force up a hill to the right and attempted to head off the British.[65]

Initially, Colonel Mawhood was not aware of General Mercer's troops; his concern was with Sullivan's column. Major Wilkinson recalled,

> *When Colonel Mawhood...discovered the head of* [Sullivan's] *column he did not perceive General Mercer, who was marching up the creek near its left bank, and taking us* [Sullivan's column] *for a light party, as the ground concealed our numbers, he determined to retrograde and cut us up; nor had General Mercer any suspicion of the proximity of Mawhood's corps, until he recrossed Stoney brook, when a mutual discovery was made at less than 500 yards distance, and the respective corps then endeavoured to get possession of the high ground on their right.*[66]

Prior to discovering Mercer's brigade, Mawhood ordered the bulk of the 55th Regiment, which was in the rear of his column, to rejoin the 40th Regiment in Princeton. Mawhood led his remaining force, several hundred strong, against Mercer. General Mercer's men collided with Mawhood's advance party in William Clark's orchard. A soldier in Mercer's detachment, known to history only as Sergeant R, described what happened:

---

[64] Ibid, 20.
[65] Ibid.
[66] Wilkinson, *Memoirs of My Own Times*, Vol. 1, 142.

As we were descending a hill through an orchard, a party of the enemy who were entrenched behind a bank and fence, rose and fired upon us. Their first shot passed over our heads cutting the limbs of the trees under which we were marching. At this moment we were ordered to wheel...We formed, advanced and fired upon the enemy. They retreated eight rods to their packs, which were laid in a line. I advanced to the fence of the opposite side of the ditch which the enemy had just left, fell on one knee and loaded my musket with ball and buckshot. Our fire was destructive; their ranks grew thin and the victory seemed nearly complete, when the British were reinforced.[67]

Mercer's men had pushed Mawhood's dismounted dragoons rearward, but now they faced Mawhood's whole force. Lieutenant James McMichael of the Pennsylvania rifle battalion recalled the encounter:

Gen. Mercer, with 100 Pennsylvania riflemen and 20 Virginians, was detached to the front to bring on the attack. The enemy then consisting of 500 [actually closer to 300] *paraded in an open field in battle array. We boldly marched to within 25 yards of them, and then commenced the attack, which was very hot. We kept up an incessant fire until it came to pushing bayonets, when we were ordered to retreat.*[68]

---

[67] Sergeant R, "The Battle of Princeton," *The Pennsylvania Magazine of History and Biography,* Vol. 20, No. 1, 517.
[68] James McMichael, "The Diary of Lt. James McMichael of the Pennsylvania Line, 1776-1778," *The Pennsylvania Magazine of History and Biography*, Vol. 16, no. 2, (1892), 141.

Despite being reinforced by the rest of his brigade, many of Mercer's men lacked bayonets and gave way. Sergeant R noted,

> *Many of our brave men had fallen, and we were unable to withstand such superior numbers of fresh troops. I soon heard General Mercer command in a tone of distress, 'Retreat'!*[69]

Mercer's men broke and fled to the rear. They abandoned two cannon and their commander, who was struck down and mortally wounded by British bayonets. Another Virginia officer struck down in the fight was Lieutenant Bartholomew Yates, of the 1st Virginia. A vivid account of his fall appeared in the *Virginia Gazette* a month after the battle.

> *In the action...* [Lieutenant Yates] *received a wound in his side which brought him to the ground. Upon seeing the enemy advance toward him, he begged for quarter; a British soldier stopped, and after deliberately loading his musket by his side, shot him through the breast. Finding that he was still alive, he stabbed in thirteen places with his bayonet; the poor youth was all the while crying for mercy.*[70]

---

[69] Sergeant R, "The Battle of Princeton," *The Pennsylvania Magazine of History and Biography,* Vol. 20, No. 1, 517.

[70] Dixon and Hunter, "Extracts of Letters from Princeton, January 31, 1777," *Virginia Gazette,* 6.

Captain John Fleming, who commanded the 1st Virginia troops on the field at Princeton that day, also fell in the battle.[71]

Help arrived for Mercer's scattered troops in the form of General Cadwalader's militiamen and a two cannon battery under Captain William Moulder.[72] Captain Thomas Rodney of Delaware noted that the appearance of Cadwalader's troops on the field momentarily checked the British advance:

> *Gen. Cadwalder's Philadelphia Brigade came up and the enemy checked by their appearance took post behind a fence and a ditch in front of* [William Clark's farm] *buildings…and so extended themselves that every man could load and fire incessantly.*[73]

With British cannon and musketry pouring in on them, Cadwalader's inexperienced troops grew jumpy. Yet, according to Captain Rodney, *"Gen. Cadwalader led up the head of the column with the greatest bravery to within 50 yards of the enemy."*[74] This was too much for many of Cadwalader's men, who broke to the rear. The crew of the two American cannon and a detachment of intrepid infantry held their ground, however, and continued to fire.[75] Captain Thomas Rodney commanded the stubborn infantrymen and recalled,

---

[71] Chase, ed., "General Washington to John Hancock, January 5, 1777," *The Papers of George Washington Revolutionary War Series,* Vol. 7, 521.

[72] Charles Wilson Peale, "Journal of Charles Wilson Peale," *Pennsylvania Magazine of History and Biography,* Vol. 38, (Philadelphia: The Historical Society of Pennsylvania, 1914), 281.

[73] Rodney, ed., *Diary of Captain Thomas Rodney,* 34.

[74] Rodney, ed., *Diary of Captain Thomas Rodney,* 34-35.

[75] Ibid. 35.

> *We...took position behind some stacks just to the left of the artillery; and about 30 of the Philadelphia Infantry were under cover of a house on our left and a little in the rear. About 15 of my men came to this post, but I could not keep them all there, for the enemies fire was dreadful....From these stacks and buildings we, with the two pieces of artillery kept up a continuous fire on the enemy; and in all probability it was this circumstance that prevented the enemy from advancing, for they could not tell the number we had posted behind these covers and were afraid to attempt passing them; but if they had known how few there were they might easily have advanced while the two brigades were in confusion and routed the whole body for it was a long time before they could be reorganized again, and indeed many, that were panic struck, ran quite off.*[76]

The determined stand of this small group of Americans not only allowed Cadwalader's and Mercer's brigades to reform, but also allowed reinforcements from the rear of General Sullivan's column to take a strong position on the battlefield. They were accompanied by General Washington who strenuously rallied Cadwalader's and Mercer's men. Sergeant R observed Washington's efforts:

> *Washington appeared in front of the American army, riding towards those of us who were retreating, and exclaimed 'Parade with us, my brave fellows, there is*

---

[76] Ibid., 35-36.

but a handful of the enemy, and we will have them directly.'[77]

The effect of his appeal was electric. Sergeant R recalled, "*I immediately joined the main body, and marched over the ground again.*"[78] Washington led the restored American line, which significantly outflanked the British, towards Mawhood's troops. The British momentarily stood firm and then began an orderly retreat. When Colonel Edward Hand's Pennsylvania riflemen moved against their left flank, the British retreat turned into a rout. Major Wilkinson noted that, "*the riflemen were…the first in pursuit, and in fact took the greatest part of the prisoners.*"[79] They were urged on by General Washington who gleefully exclaimed, "*It's a fine fox chase, boys!*"[80]

Most of the fleeing British took a circular route to Trenton, but some withdrew towards Princeton where they found the bulk of General Sullivan's column pushing the 40th and 55th Regiments. Many of the fugitives sought shelter in Nassau Hall, a large brick building in town. A blast of artillery quickly convinced them to surrender, however, and Washington's victory was complete. Ensign Beale of the 4th Virginia Regiment and his fellow Virginians in General Stephen's brigade saw little fighting in the day, but did pursue and capture over a score of British soldiers and seven officers near the college.[81]

---

[77] Sergeant R, "The Battle of Princeton," *The Pennsylvania Magazine of History and Biography,* Vol. 20, No. 1, 517.
[78] Ibid.
[79] Wilkinson, *Memoirs of My Own Times,* Vol. 1, 145.
[80] Ibid.
[81] Ryan, "Robert Beale Memoirs," *A Salute to Courage: The American Revolution as Seen Through Wartime Writings of Officers of the Continental Army and Navy,* 57.

At the cost of less than forty men killed, including Brigadier-General Hugh Mercer, Captain John Fleming and Lieutenant Bartholomew Yates of the 1st Virginia Regiment, and another forty wounded, Washington's army had inflicted a second stunning defeat on the British. General Howe's losses in killed, wounded, and captured numbered between 400 to 500 men.[82]

General Washington was tempted to stage one more daring act, an assault on the vital British supply depot at New Brunswick. He decided against it, however, because his men were exhausted and in no condition to face Cornwallis, who was rapidly marching eastward from Trenton. Reluctantly, Washington headed north, towards Morristown, and the safety of New Jersey's Watchung Mountains, to await reinforcements.

---

[82] Fischer, *Washington's Crossing*, 414-415.
  Note: General Mercer died of his wounds a few days after the battle.

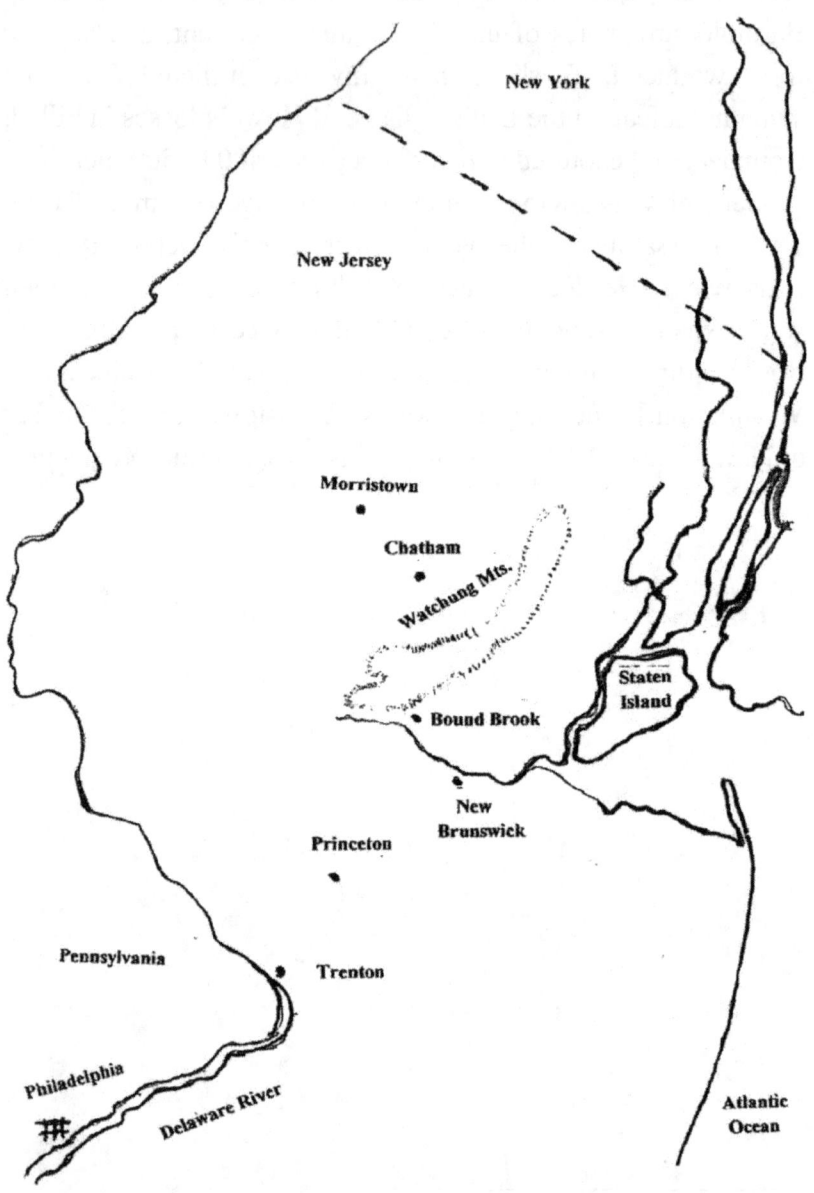

Chapter Seven

Virginia's Continental Line Grows

Winter – Summer 1777

Prior to the stunning American victories at Trenton and Princeton, Congress, unnerved by American defeats at Fort Washington and Fort Lee in November, authorized additional continental forces beyond the 88 infantry battalions it had agreed to in September. Virginia's continental artillerists who were still in Virginia, were dramatically expanded from two companies to a full artillery regiment on November 26, 1776.

The core of the artillery regiment were the two continental artillery companies raised in Virginia in the spring of 1776 that had served so well at Gwynn's Island. Captain Charles Harrison, who had assumed command of the company upon the death of Captain Dohicky Arundel at Gwynn's Island, was elevated to the rank of colonel of the regiment.

Harrison's regiment was comprised of ten artillery companies, each with a captain, three lieutenants, one sergeant, four bombardiers, eight gunners, four corporals, and forty eight matrosses.[1] Each company could fully man four artillery pieces and likely more if necessary.

Edward Carrington, who had served with Harrison at Gwynn's Island, was appointed Lieutenant Colonel of the artillery regiment and Christian Homer was appointed major

---

[1] Ford, ed., "Proceedings of the Continental Congress, November 26, 1776," *Journal of the Continental Congress*, Vol. 6, 981.

of the unit.² The new artillery companies were raised over the spring of 1777 and, like the companies before them, posted throughout Virginia to man artillery batteries at strategic locations. They were not ordered north to join the main army with General Washington until the spring of 1778.

Light Dragoons

Months before Virginia's continental artillery was expanded, in the summer of 1776, the 5th Virginia Convention authorized the formation of six troops (detachments) of provincial cavalry to help defend Virginia.³ Each troop of horse consisted of 30 rank and file privates, 3 corporals, and a trumpeter as well as a cornet (ensign), lieutenant, and captain.⁴ Thirty-four year old Theodorick Bland was chosen to command the 1st troop of cavalry and was therefore the ranking captain of the six troops of dragoons.⁵ Although he had little military experience, Bland's family connections helped overcome whatever concerns the delegates may have had.

The captains for the next three cavalry troops, Benjamin Temple, John Jamison, and Llewellin Jones, were all former captains of minute companies.⁶ Twenty year old Henry Lee was selected to command the 5th troop. Destined to be

---

² Ford, ed., "Proceedings of the Continental Congress, November 30, 1776," *Journal of the Continental Congress*, Vol. 6, 995.
³ Tarter and Scribner, eds., "Proceedings of the 5th Virginia Convention, May 20, and June 7, 1776," *Revolutionary Virginia, The Road to Independence*, Vol. 7 Part 1, 194, 390-395.
⁴ Purdie, "June 14, 1776," *Virginia Gazette*, Postscript, 4.
⁵ Tarter and Scribner, eds., "Proceedings of the 5th Virginia Convention, June 13, 1776," *Revolutionary Virginia, The Road to Independence*, Vol. 7, Part 2, 474-75.
⁶ Ibid., 475.

remembered as Light Horse Harry Lee for his bold service in the war, he likely was appointed because of his prominent father and family. John Nelson, also a member of a prominent Virginia family, commanded the sixth troop of horse.[7]

The Convention originally expected each cavalryman to provide their own horse, arms, and accoutrements, but this policy quickly changed when few troopers volunteered under such terms. To encourage enlistments the Convention announced in late June that,

> *Each trooper shall be furnished with the following arms and accoutrements...a carbine with bucket and straps, a pair of horseman's pistols and holsters, a tomahawk, a spear, and a good saddle.... And be it further ordained, that instead of the corporals, trumpeters, and private troopers, furnishing their own horses, arms, and accoutrements...the said horses, arms, and accoutrements, shall be furnished at the expense of the publick.*[8]

To compensate for the significant expense of this new policy, the convention reduced the pay of the troopers. It also extended the enlistments of the troopers to December 1, 1778.[9]

As the Virginia dragoons were initially not placed on the continental establishment by Congress, they did not ride north to join General Washington when the 1st, 3rd, 4th, 5th and 6th Virginia Regiments marched north in the fall. They spent the fall training in Williamsburg, where barracks and stables to

---

[7] Ibid.
[8] Purdie, "June 28, 1776," *Virginia Gazette*, 1.
[9] Ibid.

accommodate one hundred horses (nearly half of the entire force) were constructed. [10]

In late November, Congress requested that Governor Patrick Henry send the Virginia dragoons northward to join Washington's army.[11] They were placed on continental establishment in mid-January, retroactive to November 25, 1776 and began arriving in camp that same month.[12]

Congress also authorized the formation of three more regiments of light dragoons in late 1776. The 3rd Continental Dragoon Regiment, under the command of Colonel George Baylor of Virginia, who had served as one of General Washington's aides, consisted of six troops (detachments of about 30 horsemen) from Virginia and Maryland.[13] The first of these dragoons joined the main army in New Jersey in the spring.

Congress also ordered the continental regiments still in Virginia northward to reinforce General Washington. The 9th Virginia on the Eastern Shore was the first to be ordered north on November 23, followed a month later by the 2nd and 7th Virginia Regiments.[14] The 8th Virginia, significantly weakened by its summer deployment in Georgia, a deployment from

---

[10] H.R. McIlwaine, ed., "September 26, 1776," *Journals of the Council of the State of Virginia,* Vol. 1, (Richmond, 1931), 175.

[11] Ford, ed., "Proceedings of the Continental Congress, November 25, 1776," *Journal of the Continental Congress*, Vol. 6, 980.

[12] Ford, ed., "Proceedings of the Continental Congress, January 14, 1777," *Journal of the Continental Congress*, Vol. 7, 34.

[13] Sanchez-Saavedra, *A Guide to Virginia Military Organizations in the American Revolution, 1774-1787*, 104-105.

[14] Ford., ed., "Proceeding of the Continental Congress, November 23, and December 27, 1776," *Journals of the Continental Congress*, Vol. 6, 976, 1044.

which a portion of the regiment had yet to return to Virginia, was ordered north in January.[15]

10th – 15th Virginia Regiments

In response to Congress's call in September for additional continental regiments, Virginia's leaders went to work in October to raise six more continental regiments. Like the previous nine regiments, they were to consist of ten companies of 68 rank and file plus officers.[16] The legislature designated the number of companies each county was to raise and left it to county leaders to appoint company grade officers based on their success at recruitment.[17] Regimental officers were nominated by the legislature and approved by the Congress.

The new troops received a Congressional bounty of $20 and all of Virginia's continentals were to receive an annual clothing bounty that amounted to two suits of clothes as well as a land bounty of one hundred acres at the end of the war.[18] Officers received significantly more land for their service.

Recruitment commenced in November and continued into the next year. Enthusiasm for the war had diminished considerably in the months following the Declaration of Independence and the dire reports about Washington's army in December only added to the challenge of recruiting new troops.

The situation improved in January upon news of General Washington's victories at Trenton and Princeton. Nicholas

---

[15] Ford., ed., "Proceedings of the Continental Congress, January 21, 1777," *Journals of the Continental Congress*, Vol. 7, 52.
[16] Hening, ed., "An Act for raising six additional battalions of infantry on the continental establishment, October, 1776," *Hening's Statutes at Large...*, Vol. 9, 179-180, 183.
[17] Ibid., 180-182.
[18] Ibid., 179.

Cresswell, a loyal Englishman stranded in Virginia at the start of the war, noted the impact of the news about Trenton upon rebel recruitment.

> *News that Washington had taken 760 Hessian prisoners at Trenton...is confirmed.... The minds of the people are much altered. A few days ago they had given up the cause for lost. Their late success have turned the scale and now they are all liberty mad again. Their Recruiting parties could not get a man...and now the men are coming in by companies. Confound the [Hessians]...This has given them new spirits, got them fresh succours, and will prolong the war, perhaps for two years.*[19]

In mid-February, after several months of recruiting, the newly raised companies were assigned to the six additional continental regiments and regimental officers formally appointed.

Tenth Virginia Regiment of 1777

Colonel Edward Stevens of Culpeper County was appointed to command the 10th Virginia Regiment. He had formerly served as lieutenant colonel of the Culpeper Minute Battalion in 1775 and had seen action at the Battle of Great Bridge. His second in command was Lieutenant Colonel Lewis Willis of Spotsylvania County and George Nicholas, formerly a

---

[19] Nicholas Cresswell, "January 7, 1777," *The Journal of Nicholas Cresswell*, (New York: The Dial Press, 1924,) 179-180.

captain in the 2nd Virginia Regiment from Williamsburg, was appointed Major of the regiment.[20]

The ten companies of troops that comprised the regiment were raised in the counties of Fairfax (Captain Thomas West), Culpeper (Captain John Gillison), Amherst (Captain James Franklin), Orange (Captain Richard Stevens), Spotsylvania (Captain John Spotswood), Cumberland (Captain Tarleton Woodson), Caroline and King George (Captain David Laird), Augusta (Captain John Symmes) Stafford (Captain John Mountjoy) and Fauquier (Captain Clough Shelton).[21]

Eleventh Virginia Regiment of 1777

Colonel Daniel Morgan of Frederick County, freshly exchanged and promoted for his heroism at Quebec, commanded the 11th Virginia Regiment. His second in command was also a veteran of Quebec, Lieutenant Colonel Christian Febiger of Denmark as was Major William Heth of Frederick County.[22]

The 11th Virginia stood out from the other Virginia regiments because half of its troop strength had already been raised as rifle companies in Colonel Moses Rawling's Virginia/Maryland Rifle Corps. Unfortunately, four of these five rifle companies were captured at Fort Washington in November, leaving just Captain William Blackwell's rifle

---

[20] H.R. McIlwaine, ed., "Proceedings of the Council of State of Virginia, February 12, 1777," *Journals of the Council of State of Virginia*, Vol. 1, 339-340.

[21] Sanchez-Saavedra, *A Guide to Virginia Military Organizations in the American Revolution, 1774-1787*, 62-63.

[22] McIlwaine, ed., "Proceedings of the Council of State of Virginia, February 12, 1777," *Journals of the Council of State of Virginia*, Vol. 1, 339-340.

company from Fauquier County intact. One of the lieutenants in this company was John Marshall, future Chief Justice of the Supreme Court.[23]

The remaining companies assigned to the 11th Virginia were raised in three counties. Prince William County (Captain Charles Gallihue), provided one company while Loudoun County (Captain William Smith and Captain William Johnson) and Frederick County (Captain Peter Bruin and Captain Charles Porterfield) raised two companies each. It is likely that the troops from Frederick County were armed with rifles while those from Loudoun and Prince William carried muskets.[24]

Twelfth Virginia Regiment of 1777

Colonel James Wood of Frederick County commanded the 12th Virginia Regiment. John Neville of Augusta County was appointed Lieutenant Colonel and George Lyne of King and Queen County Major of the unit.[25]

The 12th regiment was similar to the 11th Virginia in that it had a large number of rifle companies attached to it. Five of its ten companies were raised in the frontier during the summer of 1776 to protect vulnerable settlers along the Ohio River. They were commanded by Captains Andrew Waggener, Michael Bowyer, Matthew Arbuckle, William McKee, and Benjamin

---

[23] Lee C. Bollinger, ed., *The Events of My Life: An Autobiographical Sketch by John Marshall*, (Ann Arbor, MI & Washington, D.C.: Clements Library, University of Michigan and Supreme Court Historical Society, 2001), 15.

[24] Sanchez-Saavedra, *A Guide to Virginia Military Organizations in the American Revolution, 1774-1787*, 65.

[25] McIlwaine, ed., "Proceedings of the Council of State of Virginia, February 12, 1777," *Journals of the Council of State of Virginia*, Vol. 1, 339-340.

Casey.[26] Most of the regiment's remaining companies were also raised in the western counties including Dunmore (Captain Jonathan Langdon), Berkeley (Captain Joseph Mitchell), Hampshire (Captain Stephen Ashby), and Botetourt (Captain Thomas Bowyer). The only Virginia county east of the Blue Ridge Mountains to send a company to the 12th Virginia Regiment was Prince Edward County (Captain Rowland Madison).[27]

Thirteenth Virginia Regiment of 1777

Colonel William Russell of Fincastle County commanded the 13th Virginia. He was assisted by Lieutenant Colonel John Gibson and Major Charles Simms, both of West Augusta County.[28]

This regiment was comprised completely of frontier riflemen from the distant Virginia counties of Yohogania, Monongalia, and Ohio and was principally envisioned as a frontier garrison regiment.[29]

---

[26] Ibid, and Sanchez-Saavedra, *A Guide to Virginia Military Organizations in the American Revolution, 1774-1787*, 67.
[27] Sanchez-Saavedra, *A Guide to Virginia Military Organizations in the American Revolution, 1774-1787*, 67-68.
[28] McIlwaine, ed., "Proceedings of the Council of State of Virginia, February 12, 1777," *Journals of the Council of State of Virginia*, Vol. 1, 339.
[29] Ibid.

Fourteenth Virginia Regiment of 1777

Colonel Charles Lewis of Albemarle County commanded the 14th Virginia Regiment. He was joined by Lieutenant Colonel Richard Kidder Meade of Prince George County and Major Abraham Buford of Culpeper County.[30]

Men from twelve counties were raised to fill the ranks of this regiment. The counties included, Albemarle, Louisa, Bedford, Pittsylvania, Fincastle, Goochland, Halifax, Hanover, Dinwiddie, Prince George, Lunenburg, and Charlotte.[31]

Fifteenth Virginia Regiment of 1777

Colonel David Mason of Sussex County commanded the 15th Regiment. His Lieutenant Colonel was James Innes of Williamsburg and Holt Richeson of King William County served as Major.[32]

Evidence of the increased difficulty the state had in raising the additional continental regiments is found in the number of counties it took to complete the last few regiments. The 15th Regiment drew troops from fourteen counties, including, Princess Anne, Nansemond, King William, Richmond, Westmoreland, Northumberland, Isle of Wight, South Hampton, Surry, Brunswick, Amelia, Norfolk, and Chesterfield, while the 14th Regiment consisted of troops from twelve counties.[33]

---

[30] Ibid.
[31] Ibid.
[32] Ibid.
[33] Ibid., 339-340.

Additional Continental Battalions

In late December of 1776, prior to General Washington's success at Trenton, when things still looked bleak for the American cause, the Continental Congress granted the American commander-in-chief the authority to raise another sixteen additional continental battalions beyond the 88 infantry battalions it authorized in the fall. Washington was also given authority to appoint the officers of these battalions and manage the recruitment of the troops for them (on the same terms as the other continental battalions). [34] Unlike the 88 continental regiments authorized in the fall, however, the recruits for the additional regiments were not to be raised solely in one state per regiment. The officers in command of these additional regiments has authority to, *"enlist in any of the United States of America, all such able bodied-freemen as are willing – and able to enlist...."*[35]

One of the sixteen additional regiments was commanded by and consisted of Virginians, not solely, but in large part. Colonel William Grayson of Prince William County was selected by General Washington to command an additional battalion. A former aide-de-camp to Washington, Grayson struggled in the winter and spring of 1777 to recruit men in both Virginia and Maryland for his new regiment. Aided by Lieutenant Colonel Levin Powell of Loudoun County and Major David Ross of Maryland, Grayson's Additional

---

[34] Ford, ed., "Proceedings of the Continental Congress, December 27, 1776," *Journals of the Continental Congress*, Vol. 6, 1044.

[35] Frank E. Grizzard Jr., ed., "Circular Recruiting Instructions to the Colonels of the Sixteen Additional Continental Regiments, January 12-27, 1777," *The Papers of George Washington, Revolutionary War Series*, Vol. 8, (Charlottesville: University Press of Virginia, 44.

Regiment formed ten companies commanded by captains, Thomas Tripplett, Francis Willis, John Willis, Cleon Moore, Granville Smith, Hebard Smallwood, James McGuire, Strother Jones, and Peter Grant.[36]

Colonel Nathaniel Gist, an experienced Virginia frontiersman who had served under Washington in the French and Indian War, was instructed to raise four companies of rangers from the frontier for continental service with the main army.[37] On February 5th, 1777, General Washington informed Congress that he had appointed Gist to command a regiment, *"to be raised upon the Frontiers of Virginia & Carolina."*[38] Whether this meant that Gist was given authority to raise more than four companies of rangers is unclear, but he did receive instructions to, *"bring a Company or two of Cherokee Indians,"* who Washington wished to serve as scouts (as well as hostages to insure the Cherokee would not join the British) to New Jersey with him.[39]

Winter Camp in New Jersey 1777

While recruitment of the six new continental regiments progressed slowly over the winter in Virginia, the commonwealth's light cavalry reached camp in New Jersey in mid-January. They found the situation of the American army

---

[36] Sanchez-Saavedra, *A Guide to Virginia Military Organizations in the American Revolution, 1774-1787*, 73-74.

[37] Grizzard Jr., ed., "General Washington to Colonel Nathaniel Gist, January 13, 1777," *The Papers of George Washington, Revolutionary War Series,* Vol. 8, 57.

[38] Grizzard Jr., ed., "General Washington to John Hancock, February 5, 1777," *The Papers of George Washington, Revolutionary War Series,* Vol. 8, 249.

[39] Ibid.

moderately improved from the dark days of December. Washington's decisive actions at Trenton and Princeton caused the British to pull back from western and central New Jersey and although they still held a chain of strong posts in the eastern part of the state, their reach across New Jersey and threat to the American army had diminished significantly.

The five Virginia continental regiments with General Washington numbered less than a thousand officers and men fit for duty in late December. All of the regiments were significantly understrength, with nearly as many men absent from illness as there were present in camp.[40] General Washington's decision in mid-January to send Virginia officers back to Virginia to raise new recruits for the weakened regiments suggests that these units were still significantly understrength at the start of the new year.[41]

Although the size of the American army in the mountains of New Jersey was still alarmingly small, General Washington kept pressure on the British by harassing their foraging parties whenever he could. Detachments were posted along the Watchtung Mountains to screen the main American encampment at Morristown and strike at enemy patrols and forage parties whenever the opportunity presented itself.

When the Virginia cavalry reached Philadelphia in January they were briefly assigned to a detachment under the command of General William Alexander (Lord Stirling) of New Jersey.

---

[40] Lesser, ed., "General Return of the Army of the United States... December 22, 1776," *The Sinews of Independence: Monthly Strength Reports of the Continental Army*, 43.

[41] Grizzard Jr., ed., "General Washington to General Adam Stephen, January 13, 1777," *The Papers of George Washington, Revolutionary War Series,* Vol. 8, 62.

General Stirling was posted at Basking Ridge and was under orders to,

> *Harass and annoy the Enemy by keeping Scouting parties constantly (or as frequently as possible) around their Quarters.*[42]

The Virginia cavalry was ideally suited for such activities and in all probability served in many of these scouting parties. They also participated in a few larger operations. In early February, Major Bland, who had been promoted by the Virginia government back in June, led a detachment of cavalry in a large foraging expedition to, *"remove out of* [the enemy's] *reach all of the Horses, Waggons & fat Cattle…lying between Quibble Town & the Sound, eastward; approaching as near the Enemy as you can in safety."*[43] Major Bland described the outcome of the expedition in a letter to his wife.

> *I was two days ago with part of my regiment, and a body of troops under the command of General Sullivan, on a foraging party…one or two of the light horse fired a shot or two at a small party of the enemy; a party of foot marched up to attack them, but they retreated and left us the field without the least damage done to either side. We brought off five or six hundred cattle, and about as many sheep, belonging to tories.*[44]

---

[42] Grizzard Jr., ed., "Orders to Major General Stirling, February 4, 1777," *The Papers of George Washington, Revolutionary War Series*, Vol. 8, 245.

[43] Grizzard Jr., ed., "General Washington to General Sullivan, February 3, 1777," *The Papers of George Washington, Revolutionary War Series*, Vol. 8, 237.

[44] Charles Campbell, ed., "Theodorick Bland Jr. to his lady," *The Bland*

The Virginia dragoons were not the only Virginians harassing the enemy in the winter of 1777. In late January, Colonel Mordecai Buckner of the 6th Virginia Regiment led approximately 400 men against two British regiments escorting wagons from Brunswick to Amboy. Lieutenant Colonel Josiah Parker of the 5th Virginia commanded the American advance guard, which did the bulk of the fighting. General Washington informed Congress, based on the reports he received after the fight, that, *"Our advanced party...engaged them with great Bravery upwards of twenty Minutes, during which time* [their commander] *was killed and the second in command mortally wounded."*[45] Washington added that the British losses were considerable (one report claiming thirty killed) while the Americans only lost two men who were captured.[46]

Sadly, the poor conduct of Colonel Buckner in the engagement led to his court martial for cowardice. He was cashiered in disgrace from the army.

Colonel Charles Scott of the 5th Virginia Regiment redeemed some honor to the Virginia ranks in early February in a heated engagement with several thousand enemy troops. An account of the fight appeared in the Virginia Gazette.

Papers: Being a Selection from the Manuscripts of Colonel Theodorick Bland Jr. of Prince George County, Virginia, (Petersburg: Edmund & Julian Ruffin, 1840), 47.

[45] Grizzard Jr., ed., "General Washington to John Hancock, January 26, 1777," *The Papers of George Washington, Revolutionary War Series,* Vol. 8, 161.

[46] Ibid. and Purdie, "Extract of a letter from an Officer of Distinction..." February 28, 1777," *Virginia Gazette,* Supplement, 1.

> *On [February 1] 3000 of the enemy...came out of Brunswick to forage. They had eight pieces of cannon. Several of our scouting parties joined, to the amount of 600 men, under command of Col. Scott of the 5th Virginia Regiment. A disposition was made to attack the enemy. Col. Scott, with 90 Virginians on the right, attacked 200 British grenadiers, and drove them to their cannon. The other parties not marching so briskly up to the attack, the colonel was engaged ten minutes by himself; and 300 fresh men being sent against him, was obliged to give way, but formed again within 300 yards of the enemy. By this time two other divisions had got up with the enemy, but superior numbers at last prevailed. Our troops retreated about a quarter of a mile, formed again, and looked the enemy in the face until they retreated. The enemy had 36 killed, whom the country people saw, and upwards of 100 wounded. We lost 3 officers and 12 privates killed, and have about as many wounded.*[47]

The officer added that adjutant William Kelly of the 5th Virginia and several men, all of whom had been carried off the field with wounds, were overtaken by the British and brutally murdered, the British troops, *"beating out their brains with barbarity exceeding that of the savages."*[48]

Although the Americans met with success in both engagements, the small size and weak condition of Washington's army limited his ability to conduct such operations on a regular basis. As a result, Washington's troops,

---

[47] Purdie, "Extract of a letter from an Officer of Distinction…" February 28, 1777," *Virginia Gazette*, Supplement, 1.
[48] Ibid.

particularly the Virginia cavalry, spent most of their time on small patrols, frequently skirmishing with enemy patrols and forage parties. Captain Johann Ewald of the German Jagers (riflemen) remarked in his diary that in a six week period from mid- February through March,

> *Nothing important happened due to the constant high snow, except for the daily skirmishing of our patrols and the continual alarms of the outposts on both sides. Scarcely a day passed when we did not have to stand under arms for hours in the deepest snow.*[49]

General Washington made a similar observation in April.

> *We have greatly harassed, & distressed the Enemy, by continually skirmishing with their Foraging Parties, and attacking their Picquet Guards.*[50]

The constant pressure that General Washington maintained on the enemy was quite an accomplishment given his vastly outnumbered army and something that Bland's Virginia light horse contributed significantly to. With three other continental cavalry regiments not yet complete or in camp, the workload on Major Bland's dragoons was exhausting. They were posted at several locations in New Jersey including Morristown, Chatham, and Bound Brook, and were constantly on duty.

---

[49] Joseph Tustin, ed., "Captain Johann Ewald," *Diary of the American War: A Hessian Journal*, (New Haven: Yale University Press, 1979), 55. Henceforth referred to as Ewald.
[50] Philander D. Chase, ed., "General Washington to Landon Carter, April 15, 1777," *The Papers of George Washington, Revolutionary War Series*, Vol. 9, (Charlottesville, University Press of Virginia, 1999), 171.

General Washington acknowledged the toll the constant patrols took on Bland's cavalry in a letter to General Alexander McDougall of New York in April. McDougall had asked Washington to divert some of the continental cavalry being raised in Connecticut to New York to help him defend the New York Highlands. Washington's reply was firm and revealing.

> *I cannot at this time spare any of the Continental Light Horse raising in Connecticut, they are much wanted here, those we have* [Bland's] *having* [been] *greatly reduced by the constant service since they joined me.*[51]

In late March, word arrived that Congress had re-designated the Virginia light cavalry. They were now the 1st Continental Light Dragoon Regiment.[52] Congress had placed the Virginia light dragoons onto the continental establishment two months earlier in mid-January, so the new unit designation was not a significant change.[53] It mostly meant promotions for a few of the officers. Theodorick Bland was promoted to colonel of the new regiment while Benjamin Temple and John Jameson were promoted to lieutenant colonel and major, respectively.[54]

---

[51] Chase, ed., "General Washington to General McDougall, April 17, 1777," *The Papers of George Washington, Revolutionary War Series,* Vol. 9, 187.

[52] Chase, ed., "General Orders, March 31, 1777, Footnote 3," *The Papers of George Washington, Revolutionary War Series,* Vol. 9, 25.

[53] Worthington C. Ford, ed., "January 14, 1777," *Journals of the Continental Congress,* Vol. 7, 34.

[54] Chase, ed., "General Orders, March 31, 1777," *The Papers of George Washington, Revolutionary War Series,* Vol. 9, 25.

Smallpox Inoculation

With thousands of new troops, including detachments from Virginia's fifteen continental regiments on the march to join Washington's army in New Jersey, the American commander was forced to grapple with a potential medical crisis. Smallpox ravaged his troops in 1776 and there was every reason to believe that it would do so again in 1777. General Washington wanted the troops inoculated, but the four-week process threatened to weaken the army too greatly. The lack of replacement clothing and proper facilities to care for the inoculated also convinced Washington in late January that inoculation was not possible.[55]

Despite the risk and challenges, however, General Washington changed his mind in early February and ordered all of the troops in camp or on their way to camp to be inoculated.[56] Virginia continentals were thus inoculated for smallpox over the spring of 1777 with the main army in New Jersey as well as at stops on the march in Philadelphia, Baltimore, Frederick, Maryland and Alexandria and Dumfries Virginia. Fortunately for General Washington and his troops, the British squandered their opportunity to strike the weakened American army and by April the danger had largely passed.

Although not all of the continental troops assigned to Virginia's sixteen continental regiments were in camp in New Jersey by the end of April, detachments from all of the

---

[55] Grizzard Jr., ed., "General Washington to William Shippen Jr., January 28, 1777," *The Papers of George Washington, Revolutionary War Series,* Vol. 8, 174.

[56] Grizzard Jr., ed., "General Washington to John Hancock, February 5, 1777," *The Papers of George Washington, Revolutionary War Series,* Vol. 8, 251.

regiments except the 13th, 14th, 15th, and Grayson's Additional Regiment, were.[57]

Washington Re-structures the Army

The large influx of new troops in the spring of 1777 provided General Washington with an opportunity to restructure the army into ten brigades. Each brigade, which ranged between 700 to 1,000 men, consisted of four or five undermanned regiments, typically from the same state. By mid-May the combined strength of these brigades surpassed 8,000.[58]

Four of Washington's ten brigades consisted of Virginia troops from its sixteen continental regiments. These brigades were commanded by four newly appointed brigadier generals from Virginia, Peter Muhlenberg, George Weedon, William Woodford, and Charles Scott.[59]

When Brigadier-General Andrew Lewis, the most senior of Virginia's brigadier-generals in the continental army, learned of these new arrangements, he was deeply troubled. By right he should have been promoted to Major General to replace General

---

[57] Chase, ed., "General Benjamin Lincoln to General Washington, April 27, 1777," *The Papers of George Washington, Revolutionary War Series,* Vol. 9, 286.

Note: General Lincoln writes to Washington that detachments from the 2, 3, 4, 5, 6, 7, 8, 9, 10 & 12 Virginia regiments, totally 533 privates, had left Boundbrook New Jersey (presumably for Morristown). Troops from the 11th Virginia had also joined the army by April. Those of the 13th, 14th, 15th, and Grayson's additional apparently had not.

[58] Chase, ed., "Arrangement and Present Strength of the Army in New Jersey, May 20, 1777," *The Papers of George Washington, Revolutionary War Series,* Vol. 9, 492-93.

[59] Ford., ed., "Proceedings of the Continental Congress, February 21, 1777," *Journals of the Continental Congress,* Vol. 7, 141.

Mercer, or at the very least, placed in command of General Muhlenberg's 1st Virginia Brigade. It appeared instead, that he was to languish in Virginia, recruiting troops for the army. The insult to General Lewis's honor and pride was more than he could bear and he resigned from the army in April.[60]

General Peter Muhlenberg's brigade was initially comprised of the 1st, 5th, 9th, and 13th Virginia Regiments. Colonel Moses Hazen's regiment of Canadian troops was attached to Muhlenberg's brigade to offset the missing 13th Virginia, which was split between Fort Pitt and the main army. The expected detachment of 13th Virginians had yet to arrive by May.

Muhlenberg's brigade reported just over a thousand men present and fit for duty on May 20th. The most experienced regiments in the brigade, the 1st and 5th Virginia Regiments, accounted for 120 men and 127 men respectively. Their service in New York and at Trenton and Princeton had taken a great toll on their numbers. The 9th Virginia accounted for 391 of the troop strength of the brigade and Colonel Hazen's regiment account for the rest, 393 men, for a total of 1031.[61]

General George Weedon's brigade consisted of the 2nd, 6th, 10th and 14th, Virginia Regiments. Its troop strength on May 20th, was 700 men from three of its four regiments. The 2nd Virginia accounted for 182 men, the 6th Virginia 223 men, and

---

[60] Francis Heitman, *Historical Register of Officers of the Continental Army During the War of the Revolution,* (Washington, D.C.: Rare Book Shop Publishing Company, 1914). 348.

[61] Chase, ed., "Enclosure: Arrangement and Present Strength of the Army in New Jersey, May 20, 1777," *The Papers of George Washington, Revolutionary War Series,* Vol. 9, 492.

the 10th Virginia, 295 men. The 14th Virginia had yet to arrive from Virginia.[62]

General William Woodford's brigade consisted of the 3rd, 7th, 11th and 15th Virginia Regiments. Its troop strength on May 20th, was 999 men from three of its four regiments. The veteran 3rd Virginia accounted for just 150 men, the 7th Virginia 472 men, and the 11th Virginia 377 men. The 15th Virginia had yet to arrive from Virginia.[63]

The last of Virginia's four brigades was commanded by General Charles Scott. Like Virginia's other three brigades, one regiment, Colonel Grayson's Additional Regiment, had yet to report to the army. General Washington attached Col. John Patton's Additional Regiment to Scott's brigade to offset the loss of Colonel Grayson's missing regiment.

General Scott's brigade totaled 712 men on May 20th. The 4th Virginia accounted for 314 men, the 8th Virginia, 157 men and the 12th Virginia just 117 men. Colonel Patton's Additional Regiment added another 124 men to the brigade.[64]

1st Continental Dragoons

The absence of the bulk of the continental cavalry (which was still being raised among the states) placed an enormous burden on Colonel Bland's Virginia horsemen. Congress had authorized the formation of four continental cavalry regiments during the winter of 1776-77, but only one, Bland's 1st Continental Dragoons, had joined Washington's army at full strength. Colonel Stephen Moylan's 4th Continental Dragoons

---

[62] Ibid.
[63] Ibid.
[64] Ibid.

(raised in the mid-Atlantic region and Virginia) and Colonel George Baylor's 3rd Continental Dragoons (raised in Virginia and Maryland) each had sent a troop of cavalry to Washington by June, but most of the dragoons in these regiments, along with Colonel Elisha Sheldon's entire 2nd Continental Dragoon Regiment (raised in Connecticut, Massachusetts and New Jersey) had yet to arrive in camp.[65]

Scattered among numerous American posts in New Jersey and constantly on patrol, the endless duty wore down the Colonel Bland's cavalrymen and their horses to such a point that General Washington agreed to temporarily withdraw the regiment from service in July. Washington discussed his decision in a letter to General John Sullivan.

> *The Virginia Regiment of Horse had been so detached the whole Winter that I could not deny Colo. Bland his request to draw them together that they may be properly equipped, which they have never yet been.*[66]

To replace Bland's exhausted dragoons, Washington ordered Colonel Elisha Sheldon, who was still organizing his cavalry regiment in Connecticut, to hurry his men along.

---

[65] Chase, ed. "General Washington to Colonel Baylor, May 17, 1777," and "General Washington to Colonel Sheldon, May 24, 1777," *The Papers of George Washington, Revolutionary War Series,* Vol. 9, 448, 521, and Frank E. Grizzard, Jr., ed., "General Washington to General Schuyler, June 16, 1777," *The Papers of George Washington, Revolutionary War Series,* Vol. 10, (Charlottesville: University Press of Virginia, 2000), 54.

[66] Chase, ed., "General Washington to General Sullivan, June 7, 1777," *The Papers of George Washington, Revolutionary War Series,* Vol. 9, 639.

> *The Virginia regt of Light Horse have been so worn down by hard service that* [unless] *they are reliev'd of part of their Duty they will be totally unfit for service of any kind. I therefore desire that you will send on ev'ry man of your Regt that is cloathed & Mounted and that have had the small Pox.*[67]

Colonel Bland's exhausted cavalry regiment was temporarily removed from active service in July to rest and recover from the hard winter and spring.[68] A top priority was to allow the regiment to, "*get forage to recruit their horses.*"[69]

Morgan's Rifle Corps

Although Colonel Bland's 1st Continental dragoons were indispensable to General Washington in the spring of 1777, they could not replace the need for a corps of light infantry that Washington desired to counter the enemy's light infantry.

The British used their light infantry (who were typically selected for their mobility) as flankers and skirmishers. They often combined these soldiers with German Jagers (riflemen) and British dragoons for reconnaissance activities.

General Washington largely relied on militia troops for such duty in 1776 and found them ineffective. Washington's orders for June 1st, 1777, suggest that he had settled on a replacement for the militia.

---

[67] Chase, ed., "General Washington to Colonel Sheldon, June 9, 1777," *The Papers of George Washington, Revolutionary War Series,* Vol. 9, 652-53.

[68] Frank E. Grizzard, Jr., ed., "General Orders, July 1, 1777," *The Papers of George Washington, Revolutionary War Series*, Vol. 10, (Charlottesville: University Press of Virginia, 2000), 162.

[69] Ibid.

> *The commanding officer of every Corps is to make a report early to-morrow morning...of the number of Rifle-men under his command—In doing which, he is to include none but such as are known to be perfectly skilled in the use of these guns, and who are known to be active and orderly in their behaviour.*[70]

Two weeks later, on June 13th, Washington formed a new rifle corps and appointed Colonel Daniel Morgan, the commander of the 11th Virginia Regiment, to its command. Washington informed Colonel Morgan that,

> *The corps of Rangers newly formed and under your command, are to be considered as a body of light infantry, and are to act as such, for which reason they will be exempted from the common duties of the line.*[71]

An apparent shortage of rifles caused Washington to order,

> *Such rifles as belong to the States, in the different brigades, to be immediately exchanged with Col. Morgan for musquets...If a sufficient number of rifles (public property) can not be procured, the Brigadiers are requested to assist Col. Morgan, either by exchanging, or purchasing those that are private property.*[72]

---

[70] Chase, ed., "General Orders, June 1, 1777," *The Papers of George Washington, Revolutionary War Series,* Vol. 9, 578.

[71] Grizzard Jr., ed., "General Washington to Colonel Morgan, June 13, 1777," *The Papers of George Washington, Revolutionary War Series,* Vol. 10, 31.

[72] Chase, ed., "General Orders, June 13, 1777," *The Papers of George Washington, Revolutionary War Series,* Vol. 10, 20.

General Washington also made arrangements for the riflemen to receive spears as a defense against mounted troops. He informed Colonel Morgan that,

> *I have sent for Spears, which I expect shortly to receive and deliver you, as a defence against Horse; till you are furnished with these, take care not to be caught in such a Situation as to give them any advantage over you.*[73]

The spears arrived a week later. General Washington was pleased, but recommended a few adjustments to suit the riflemen:

> *The Spears have come to hand, and are very handy and will be useful to the Rifle Men. But they would be more conveniently carried, if they had a sling fixed to them, they should also have a spike in the but end to fix them in the ground and they would then serve as a rest for the Rifle. The Iron plates which fix the spear head to the shaft, should be at least eighteen inches long to prevent the Shaft from being cut through, with a stroke of a Horseman's Sword.*[74]

It is unclear how long the riflemen actually used the spears. No further reference to them appears after June 20th, which

---

[73] Grizzard Jr., ed., "General Washington to Colonel Morgan, June 13, 1777," *The Papers of George Washington, Revolutionary War Series*, Vol. 10, 31.

[74] Grizzard Jr., ed., "General Washington to Richard Peters, June 20, 1777," *The Papers of George Washington, Revolutionary War Series*, Vol. 10, 88.

suggests the cumbersome weapons were quickly discarded by the riflemen.

Morgan's corps consisted of riflemen from Virginia and Pennsylvania (500 in all evenly split), and the unit was immediately put to use. On the day of its formation, General Washington ordered Colonel Morgan to

> *Take post at Van Vechten's Bridge, and watch, with very small scouting parties (to avoid fatiguing your men too much...) the enemy's left flank...In case of any movement of the enemy, you are instantly to fall upon their flanks, and gall them as much as possible, taking especial care not to be surrounded, or to have your retreat to the army cut off.*[75]

The next day, General Howe sent a large detachment towards the American lines. Washington sent Morgan's rifle corps forward to skirmish with them. He described the engagement in a letter to General Sullivan:

> *The Enemy have advanced a party; said to be two thousand, as far as Van Ests Mill upon Millstone River. They have been skirmishing with Colo. Morgans Riflemen but have halted on a piece of high ground.*[76]

---

[75] Grizzard Jr., ed., "General Washington to Colonel Morgan, June 13, 1777," *The Papers of George Washington, Revolutionary War Series,* Vol. 10, 31.

[76] Grizzard Jr., ed., "General Washington to General Sullivan, June 14, 1777," *The Papers of George Washington, Revolutionary War Series,* Vol. 10, 40.

Two days later Washington re-iterated his orders to Colonel Morgan:

> *You will continue to keep out your active parties carefully watching every motion of the enemy; and have your whole body in readiness to move without confusion, and free from danger....*[77]

The British held their position for nearly a week and skirmished daily with Morgan's riflemen. When Howe finally withdrew, Morgan's rifle corps pursued them all the way to Piscataway. Washington noted that, *"In the pursuit, Colo. Morgans Rifle Men exchanged several sharp Fires with the Enemy, which it is imagined did them considerable execution.*[78] Captain Thomas Posey of Virginia commanded one of Morgan's rifle companies and provided a detailed account of the engagement in his biography.

> *The [rifle] regiment was posted in a thick wood somewhat swampy near the rode, & when the main body of the enemy passed, & the rear guard came on, Morgan ordered the regiment to attack and indeavour to cut it off. The order was promptly obeyed, & the action was warmly contested on both sides; in the course of the action Capt. Posey was ordered with his*

---

[77] Grizzard Jr., ed., "Richard Meade to Daniel Morgan, June 16, 1777," *The Papers of George Washington, Revolutionary War Series,* Vol. 10, 40.

[78] Grizzard Jr., ed., "General Washington to John Hancock, June 22, 1777," *The Papers of George Washington, Revolutionary War Series,* Vol. 10, 104-105.

> *company across a causeway, being through a considerable swamp to gain the front of the enemy which was promptly executed & a sharp conflict took place, but the light Infantry of the enemy surrounded his company, and was near cutting him off* [when] *he, perceiving his situation, ordered a well directed fire upon a particular part of the enemy, which opened a passage for him to retreat through. Through the course of the action the regiment sustained considerable loss in killed & wounded, the enemy suffered very considerably, tho' the highest loss fell upon Capt. Posey's company.*[79]

General Washington believed that Morgan's rifle corps had soundly thrashed the enemy:

> *I fancy the British Grenadiers got a pretty severe peppering yesterday by Morgan's Riffle Corp – they fought it seems a considerable time within the distance of, from twenty to forty yards…more than a hundd of them must have fallen.*[80]

Four days later, General Howe renewed his efforts to draw Washington out of his fortified lines with a sudden march towards the American left flank. This movement surprised the Americans and was nearly a disaster for the rifle corps. Captain

---

[79] Thomas Posey, "A Short Biography of the Life of Governor Thomas Posey," *Thomas Posey Papers*, Indiana Historical Society Library, Indianapolis, Indiana.

[80] Gizzard Jr., ed., "General Washington to Joseph Reed, June 23, 1777," *The Papers of George Washington, Revolutionary War Series,* Vol. 10, 115.

John Chilton of the 3rd Virginia Regiment, described what happened:

> *Col. Morgan with the Rifle Regmt. was on the Mattuchin lines at the time and our main army had come down into the Plains. The Enemy unexpectedly stole a march in the night of the 25th and had nearly surrounded Morgan before he was aware of it. He with difficulty saved his men and baggage and after a retreat, rallied his men and sustained a heavy charge until reinforced by Major Genl. Ld. Stirling, who gave them so warm a reception that they were obliged to retreat so precipitately that it had like to have become a rout. But being strongly reinforced he* [Stirling] *was obliged to retreat with the loss of 2 pieces of Artillery.*[81]

Although General Washington's advance parties had been surprised and pressed by the British, the main American position remained secure. General Howe, discouraged at his failure to draw the American army away from its fortified position and into an open engagement, withdrew to New York to develop a new plan of attack.

---

[81] Tyler, "John Chilton to his brother, June 29, 1777", *Tyler's Quarterly Historical and Genealogical Magazine*, Vol. 12, 118.

Tour of the Jerseys

On July 12, with indications that General Howe was heading up the Hudson River by ship, Washington set the American army in motion. Over the next eleven days Washington's troops made their way northward, arriving in Chester, New York on July 23. They remained their just one day, however, and hastily redirected their march south.[82]

The reason for this sudden change in direction was a report that the British fleet transporting General Howe's army was actually heading south from New York. Unsure of its ultimate destination, but convinced that his army was dangerously out of position to intercept Howe's army, Washington commenced a forced march back towards Philadelphia.

Captain John Chilton of the 3rd Virginia Regiment kept a detailed account of this march in his diary, noting that the army marched 23 miles on July 25. The troops got a very early start on July 26, striking their tents at 3:00 a.m. and marching eleven miles by 9:00 a.m. They marched a total of 25 miles that day, and, despite a late start due to rain, they marched another 21 miles the next.[83]

Such hard marching in the summer heat (seventy miles in three days) took a toll on the men. Captain Chilton recorded the hardships that were endured in his diary.

We were ordered to sit down, in the Sun no water near, to refresh ourselves no victuals to eat as the (march) of

---

[82] Tyler, ed., "Chilton Diary, July 12-23, 1777," *Tyler's Quarterly Historical and Genealogical Magazine*, Vol. 12, 284.

[83] Tyler, ed., "Chilton Diary, July 25-27, 1777," *Tyler's Quarterly Historical and Genealogical Magazine*, Vol. 12, 285.

last night was so late that nothing could be cooked, no wagons allowed to carry our Cooking Utensils, the soldiers were obliged to carry their Kettles pans &c. in their hands, Clothes and provisions on their backs.[84]

The July heat made every step miserable and the men suffered greatly for want of shoes. Chilton continued,

As our March was a forced one & the Season extremely warm the victuals became putrid by sweat & heat – the Men badly off for Shoes, many being entirely barefoot.[85]

Fortunately for the men, the pace of the march soon slowed with the regiment covering only 35 over the next three days. The 3rd Virginia arrived at the Delaware River on July 31, and experienced a mishap while crossing. One of the wagons carrying many of the regiment's tents, including Captain Chilton's, over-turned while being ferried across the river. That night the 3rd regiment was, *"obliged to take to the woods for want of Tents."*[86]

For most of August, the American army was stationed only a day's march outside Philadelphia. Once again, General Washington had a difficult time determining the intentions of General Howe. On two separate occasions, Washington concluded that Howe's naval movement to the south was actually a feint to draw the American army in that direction.

---

[84] Tyler, ed., "Chilton Diary, July 27, 1777," *Tyler's Quarterly Historical and Genealogical Magazine*, Vol. 12, 286.
[85] Ibid.
[86] Tyler, ed., "Chilton Diary, July 31, 1777," *Tyler's Quarterly Historical and Genealogical Magazine*, Vol. 12, 286.

Each time he prepared to march north however, news arrived that the British fleet was still sailing south. Unbeknownst to General Washington, the fleet had encountered contrary winds, which pushed them off course and delayed their voyage southward for weeks.

This delay confused General Washington. Although he still felt that Howe would attempt to join General Burgoyne in New York, the sporadic reports of the fleet heading south caused Washington to keep his army in Pennsylvania until he could definitely determine Howe's destination.

After long marches back and forth across New Jersey, the American army welcomed the opportunity to rest. And for most of August, that is what they did.

On August 21, General Washington held a Council of War, which determined that General Howe had likely sailed to Charleston, South Carolina and had become delayed at sea. The council advised Washington not to march to Charleston because such an attempt would be extremely injurious to the army and they could not possibly arrive in time to be of any service to South Carolina. The council recommended instead, and General Washington agreed, that the main army should march north and join the northern army in New York that was trying to stop General Burgoyne's thrust towards the Hudson River.[87]

Before the army commenced its march however, a new sighting of the British fleet placed it in a shocking position. General Howe was sailing up the Chesapeake Bay.

Washington reacted immediately, ordering the army to march to Philadelphia. On August 23, with the army just five miles outside the capital, Washington issued orders for, *"Every*

---

[87] John Reed, *Campaign to Valley Forge*, (Pioneer Press, 1965), 58-59.

Man to have clean clothes ready for the Morning, the Arms to be Furbished & brights."[88] He wanted to make a good impression upon the inhabitants of Philadelphia.

Early in the morning of August 24, the army, with sprigs of green in their hats and polished muskets in their hands, marched through Philadelphia. Camp followers (wives, families, others) along with most of the supply wagons, were divered around the city.[89] The two hour procession seemed to impress the city's inhabitants, who had lined the route to watch and cheer. One observe, John Adams of Massachusetts, wrote to his wife Abigail that, *"We have an army well appointed between us and Mr. Howe...so that I feel as secure here as if I was at Braintree."*[90] However, Adams also noted that,

Our soldiers have not yet quite the air of soldiers. They don't step exactly in time. They don't hold up their heads quite erect, nor turn out their toes so exactly as they ought...[91]

General Washington kept the army moving south towards Wilmington, Delaware. General Howe's troops began disembarking from their ships near Head of Elk, Maryland, twenty miles to the southwest.

General Washington continued past Wilmington and made preparations for battle. Just two weeks earlier he had ordered

---

[88] Philander D. Chase and Edward G. Lengel, eds., "General Orders for August 24, 1777," *The Papers of George Washington, Revolutionary War Series*, Vol. 11, (Charlottesville: University Press of Virginia, 2001), 55.
[89] Reed, 79.
[90] Ward, *Duty, Honor, or Country: General George Weedon and the American Revolution*, 96.
[91] Reed, 79

Colonel Morgan's rifle corps to march north to reinforce the American northern army in New York under General Philip Schuyler. In need of a replacement for Morgan's riflemen, Washington formed a new corps of light infantry on August 28.

> *A corps of Light Infantry is to be formed, to consist of one Field Officer, two Captain, six subalterns, eight Serjeants and 100 Rank and File from each brigade.*[92]

As four of the army's ten brigades comprised Virginia men, nearly half of this new light corps, some 450 officers and men, were likely Virginians. General William Maxwell of New Jersey was selected by General Washington to command this corps. Colonel William Crawford of Virginia served as second in command.[93] He had served as lieutenant colonel of the 5th Virginia in the spring of 1776 and was promoted to colonel of the 7th Virginia in August of that year, but resigned in March 1777, apparently to attend to personal affairs on the frontier.[94]

It is unclear in what capacity Colonel Crawford had returned to the army in the late summer of 1777, but several pension accounts of Virginians who served in Maxwell's Light Corps cite Crawford as one of the unit's commanders.[95]

---

[92] Chase and Lengel, eds., "Additional After Orders to the General Orders for August 28, 1777," *The Papers of George Washington, Revolutionary War Series*, Vol. 11, 82.

[93] Gabriel Neville, "The B Team of 1777: Maxwell's Light Infantry," *Journal of the American Revolution*, (Online: April 10, 2018).

[94] Chase and Lengel, eds., "From Joseph Reed, September 23, 1777, Footnote 4," *The Papers of George Washington, Revolutionary War Series*, Vol. 11, 307.

[95] Neville, "The B Team of 1777: Maxwell's Light Infantry," *Journal of the American Revolution*, (Online: April 10, 2018).

Lieutenant Colonel William Heth of General Woodford's brigade, Lieutenant Colonel Richard Parker of General Weedon's brigade, and Major Charles Simms of General Scott's brigade also served in the new light corps.[96] If General Washington's orders on the formation of the light corps were strictly followed, these Virginian officers were joined by 8 captains, 24 lieutenants, 32 sergeants and 400 troops, all from Virginia.

On August 30, Maxwell's corps was ordered forward to the vicinity of Cooch's Bridge, a likely route for the British. General Washington instructed Maxwell to place some of his men at the pass on the road and annoy the enemy should they attempt to march through. He added that the men should like quiet and expect the enemy to move early in the morning.[97]

Battle of Cooches Bridge

It was another three days before the British marched, and when they did, they followed the route General Washington expected, towards Iron Hill and Cooches Bridge. An advance guard, consisting of German jaegers, British light infantry, and a few dragoons, led the army.

The route to the bridge was ideally situated for the type of fighting General Maxwell desired. Aware that his detachment was greatly outnumbered, Maxwell had no intention of facing the enemy in open combat. Instead, he hoped to delay their march with a series of ambushes. His men were instructed to strike the enemy from concealed positions and when pressed,

---

[96] Ibid.,
[97] Chase and Lengel, eds., "General Washington to William Maxwell, August 30, 1777," *The Papers of George Washington, Revolutionary War Series*, Vol. 11, 95.

to fall back, reform, and hit them again. Thus, every tree, thicket, and rock along the road was a possible firing position for the Americans. British Captain John Montresor ominously described the terrain in his journal. *"The Country is close-- the woods within shot of the road, frequently in front and flank and in projecting points towards the Road."*[98]

General Howe's advance guard of jaegers and dragoons engaged General Maxwell's advance troops about 9:00 a.m. on September 3. Johann Ewald, a captain with the German Jaegers, accompanied by six dragoons, rode ahead of the British advance corps and described the initial contact.

> *I...had not gone a hundred paces from the advance guard,"* recalled Captain Ewald, *"when I received fire from a hedge, through which these six men* [the dragoons] *were all either killed or wounded. My horse, which normally was well used to fire, reared so high several times that I expected it would throw me. I cried out, "Foot jagers forward!" and advanced with them to the area from which the fire was coming...At this moment I ran into another enemy party with which I became heavily engaged. Lieutenant Colonel von Wurmb, who came with the entire Corps assisted by the light infantry, ordered the advance guard to be supported.*[99]

---

[98] John Montresor, "Journal of Captain John Montresor, September 3, 1777," *The Pennsylvania Magazine of History and Biography* Vol. 5, (Philadelphia: The Historical Society of Pennsylvania, 1881), 412.

[99] Ewald, 77.

The Americans, according to plan, fell back. *"A Continued Smart irregular fire* [ensued] *for near two miles,"* recalled Captain Montresor.[100] Sergeant Thomas Sullivan of the British 49th Regiment, attributed the, "hot fire" of the Americans to their strong position.[101]

The engagement lasted into the afternoon with the Americans fighting from tree to tree. They gradually withdrew to Iron Hill and Cooches Bridge. As the British pressed forward, they saw American troops scurrying in the woods on Iron Hill.[102] General Howe ordered the British advance guard to drive the enemy off the mountain. Captain Ewald led the way and recalled that,

> *The charge was sounded, and the enemy was attacked so severely and with such spirit by the jagers that we became masters of the mountain after a seven hour engagement.*[103]

Ewald described the fighting as intense and hand to hand.

> *The majority of the jagers came to close quarters with the enemy, and the hunting sword was used as much as the rifle.*[104]

---

[100] Montressor, 412.
[101] Thomas Sullivan, "Before and After the Battle of Brandywine: Extracts from the Journal of Sergeant Thomas Sullivan of H.M. Forty-Ninth Regiment of Foot", *The Pennsylvania Magazine of History and Biography,* Vol. 31, (Philadelphia: Historical Society of Pennsylvania, 1907), 410
[102] Ewald, 78.
[103] Ibid.
[104] Ibid.

The British also used artillery in the attack, but with little effect. One of the biggest threats to the Americans occurred when Lieutenant Colonel Robert Abercromby led a battalion of British light infantry across Christianna Creek and around the left flank of Maxwell's corps. Abercromby hoped to cut off Maxwell's route of retreat. Fortunately for the Americans, Abercromby's battalion stumbled upon an impassable swamp and was unable to complete the encirclement.

It is unclear whether the Americans were even aware of this threat to their rear. They had their hands full with the jaegers, who steadily pushed Maxwell's men off Iron Hill and towards Cooches Bridge. *"The jagers alone enjoyed the honor of driving the enemy out of his advantageous position,"* boasted Ewald years later.[105]

Although the Americans were forced from Iron Hill, they still had some fight left in them. Sergeant Sullivan observed that,

> *After a hot fire the enemy retreated towards their main body, by Iron Hill. They made a stand at the Bridge for some time, but the pursuing Corps made them quit that post also, and retire with loss.*[106]

General Maxwell withdrew from Cooches Bridge in the early afternoon and rejoined the main army, content that his troops had performed their mission admirably. They had harassed the enemy all morning and delayed their advance to

---

[105] Ibid.
[106] Sullivan, 410.

the point that, when the skirmish ended, General Howe halted his army for the rest of the day.[107]

General Washington, pleased with the conduct of the light corps, sent a report to Congress, speculating that Maxwell's detachment inflicted considerable damage on the enemy.

> *This morning,"* he wrote, *the Enemy came out with considerable force and three pieces of Artillery, against our Light advanced Corps, and after some pretty smart skirmishing obliged them to retreat, being far inferior in number and without Cannon. The loss on either side is not yet ascertained. Our's, tho not exactly known, is not very considerable; Theirs, we have reason to believe, was much greater, as some of our parties composed of expert Marksmen, had Opportunities of giving them several, close, well directed Fires, more particularly in One instant, when a body of Riflemen formed a kind of Ambuscade.*[108]

Although General Washington was confident that the Americans got the best of the British, accurate casualty figures are difficult to determine. Both sides claimed they inflicted more loss on the enemy than they sustained. It appears, however, that the losses were rather light, ranging between twenty- five to fifty men each.[109]

---

[107] Montressor, 413.

[108] Chase and Lengel, eds., "General Washington to John Hancock, September 3, 1777," *The Papers of George Washington, Revolutionary War Series,* Vol. 11, 135.

[109] Note: General Howe reported losses of 3 dead and 21 wounded. Sergeant Thomas Sullivan reported identical numbers in his journal. Captain Ewald, however, claimed losses of 11 dead and 45 wounded. The reports of American losses also vary. Captain Montresor claimed that,

Following the battle, Maxwell's Corps retreated towards the American army, which was entrenched along Red Clay Creek. They took up new positions in advance of the army and were ordered to maintain a close watch on the enemy.[110] Two days after the skirmish, General Washington gave General Maxwell the following instructions.

> *I should be glad to hear how the Enemy are situated and what they seem to be about. Send out reconnoitering parties under good intelligent officers to inspect the different parts of their Camp, and gain as exact an insight as possible into their circumstances…You should always have small advanced parties towards the Enemy's lines, about the hour of the morning you expect them to move, as it is of essential importance to us, to have the earliest intelligence of it.*[111]

The situation changed on September 7. Reports reached General Washington that the enemy had stripped itself of its excess baggage in preparation for a march. General Washington responded with similar orders.

"the rebels left about 20 dead". Major Baurmeister put that number at 30. Captain Muenchhausen claimed 41 rebels were buried by the British, including five officers. For his part, General Washington reported to Congress that, *"…we had forty killed and wounded, and as our own Men were thinly posted they must have done more damage upon a close Body then they received."*

[110] Reed, 89.

[111] Chase and Lengel, eds., "General Washington to General William Maxwell, September 5, 1777," *The Papers of George Washington Revolutionary War Series,* Vol. 11, 154.

> *The General has received a confirmation...that the enemy have disencumbered themselves of all their baggage, even to their tents, reserving only their blankets and such part of the cloathing as is absolutely necessary. This indicates a speedy & rapid movement, and points out the necessity of following the example, and ridding ourselves for a few days of every thing we can possibly dispense with...Officers should only retain their blankets, great coats, and three or four shirts of under cloaths, and the men should, besides what they have on, keep only a Blanket, and a shirt a piece, and such as have it, a great coat – All trunks, chests, boxes, other bedding, and cloathes... [are] to be sent away, 'till the elapsing of a few days shall determine whether the enemy mean an immediate attack, or not.*[112]

Another day passed before General Howe's intentions became clear. Rather than confront the Americans in a costly frontal assault of their entrenched position, General Howe marched his army north, in an attempt to gain the right flank of the Americans.

Determined to protect both his flank, and Philadelphia, General Washington raced his army northward. His destination was Chad's Ford, the likely British crossing point over Brandywine Creek. The Americans marched with urgency and arrived at the ford on September 9th.

---

[112] Chase, and Lengel, eds., "General Orders for September 7, 1777," *The Papers of George Washington, Revolutionary War Series*, Vol. 11, 167-168.

Advance elements of General Howe's army, which had a longer route to march, arrived at Kennett Square, seven miles west of Chads Ford, that same evening. General Howe, with the main portion of his army, arrived the next morning. He remained at Kennett Square on September 10th, consolidating his force and devising a plan of attack. The stage was now set for a major clash, and the fate of Philadelphia hung in the balance.

Battle of Brandywine

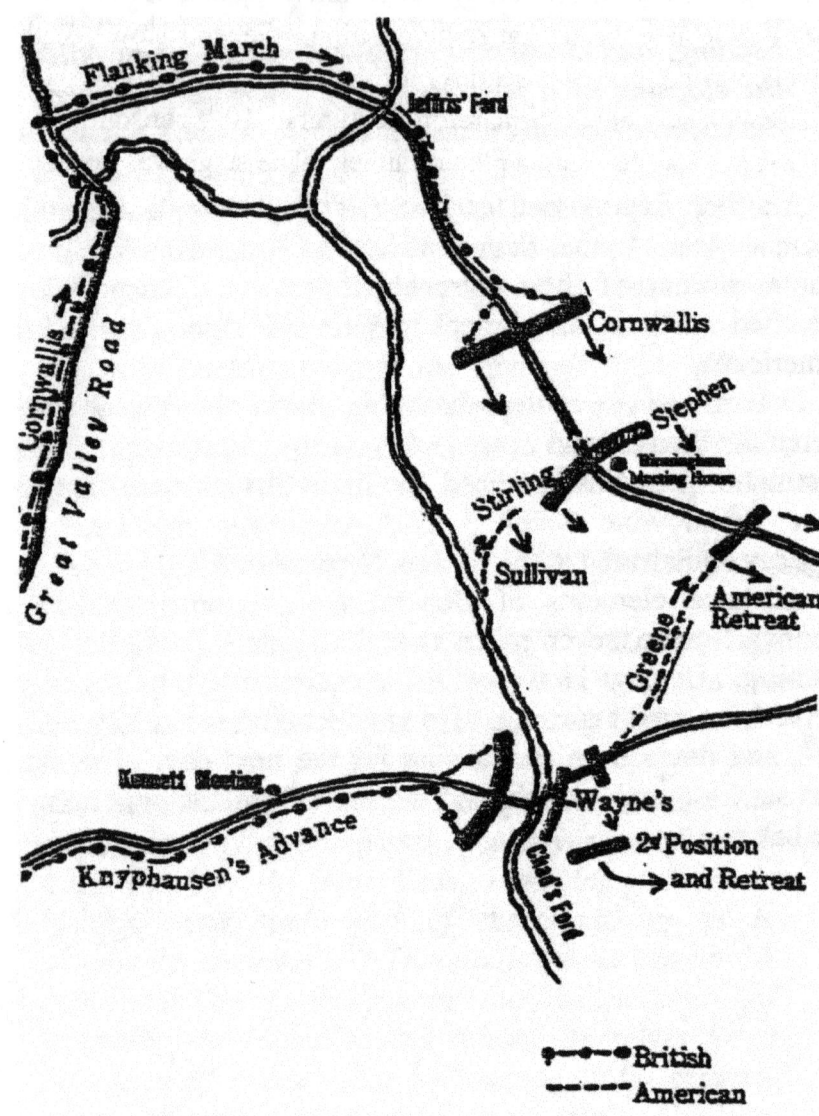

Chapter Eight

Battle of Brandywine

On the evening of September 10, 1777, approximately 30,000 troops, evenly split between two sides, prepared to do battle thirty miles west of Philadelphia. A modest creek, the Brandywine, separated the two armies, creating a barrier General Washington used to deploy his troops behind.

The center of Washington's line of battle was at Chads's Ford, one of several places along the Brandywine low enough for troops to ford. General Anthony Wayne's 2,000 Pennsylvanians defended this crossing point. General Nathanael Greene's division of 2,500 Virginians, made up of the brigades of General George Weedon and General Peter Muhlenberg, defended the ground to Wayne's left, immediately south of the ford. They were supported by General Francis Nash's Carolina Brigade (1,000 strong).[1]

General Adam Stephen's division of Virginians made up of the brigades of General William Woodford and General Charles Scott brigades, 2,100 men strong, as well as General William Alexander (Lord Stirling's) division of Pennsylvania and New Jersey troops, also 1,400 men strong, were posted in reserve on a hill overlooking Brandywine Creek, just north of Chads's Ford.[2] About a mile north of their position, further up Brandywine Creek, was General John Sullivan with a division

---

[1] Michael C. Harris, *Brandywine: A Military History of the Battle that Lost Philadelphia but Saved America, September 11, 1777*, (California: Savas Beatie, 2014), 167-168, 172.
[2] Ibid., 171-172.

of 1,800 Maryland troops.³ They guarded another ford and also policed the American right flank against any surprise by General Howe. With battle seemingly imminent, General Washington posted General Maxwell's 1,000 man light corps across Brandywine Creek to watch the enemy's movements and harass them if and when they moved against the Americans.

General Maxwell positioned a large portion of his men, as well as a few light cannon, on a ridge overlooking the west bank of the Brandywine and the road leading to the ford.⁴ He placed smaller detachments further west, towards the enemy.

General Light Horse Harry Lee, writing about the battle in his memoirs years later, wrote that,

> *Three small detachments, commanded by Lieutenant Colonels Parker, Heth, and Simms, of the Virginia line, were early in the morning posted by the brigadier* [Maxwell] *contiguous to the road, some distance in his front; and Captain Porterfield, with a company of infantry, preceded these parties with orders to deliver his fire as soon as he should meet the van of the enemy, and then fall back.*⁵

Horse patrols extended almost to Kennett Square with orders to sound the alarm when the enemy approached.

The British commenced their march towards the Americans before daybreak, on September 11th 1777.⁶ They marched in

---

³ Ibid., 170.
⁴ Samuel Smith, *The Battle of Brandywine,* (Monmouth Beach, NJ: Philip Freneau Press, 1976), 10.
⁵ Henry Lee, *The Revolutionary War Memoirs of General Henry Lee,* ed. by Robert E. Lee, (New York: Da Capo Press, 1998), 89.
⁶ Sullivan, 412.

two separate columns along two separate roads.[7] General Howe, with over 8,000 men, headed north, on a seventeen mile trek that, he hoped, would place him on the right flank of the American army.[8] General Knyphausen, with just under 7,000 men, marched east, straight towards the American army at Chad's Ford.[9] His column was a decoy, or holding force. General Howe hoped to duplicate his success at Long Island by feigning a frontal attack, and striking the Americans on their vulnerable flank. Timing and deception were key elements of the plan. General Knyphausen had to convince the Americans that his force was the main assault. First, however, Knyphausen had to deal with General Maxwell's light infantry.

Leading Knyphausen's column were the Queen's Rangers, under Captain James Wemys of the 40th Regiment, and Captain Patrick Ferguson's Rifle Corps. Ferguson's men carried a novelty to war, breech loading rifles. These weapons, developed by Captain Ferguson himself, allowed for faster and more accurate fire, and Ferguson was eager to demonstrate their usefulness in battle. The Rangers numbered 350 and the riflemen, 130.[10] A small squad of horse was also attached to the advance guard.

Knyphausen's vanguard encountered a detachment of American light horsemen less than a mile into their march. They were posted at Welch's Tavern and were nearly caught

---

[7] Bernard Uhlendorf and Edna Vosper, trans. and eds., "Letters of Major Baurmeister During the Philadelphia Campaign," *The Pennsylvania Magazine of History and Biography,* Vol. 59, (Philadelphia: Historical Society of Pennsylvania, 1935), 404. Henceforth referred to as Baurmeister.
[8] Smith, 9.
[9] Ibid.
[10] Ibid. 10.

by surprise by the British advance.[11] The Americans scurried out the back door of the tavern, escaping injury and capture. Their horses, however, did not fare so well, and were pressed into British service.[12]

The first deadly encounter between the foes occurred soon afterwards. As the British approached Kennett Meetinghouse, they were suddenly attacked by Captain Charles Porterfield's company of Virginians. Lieutenant Colonel William Heth, in a letter to Colonel Daniel Morgan, described the incident.

> *Our valuable Friend Porterfield began the action with day light – he killed (himself) the first man who fell that day – His conduct through the whole day – was such, as has acquired him the greatest Honor – A great proportion of British Officers fell by a party under his command & Capt. Waggoners (who is a brave officer) and I find it impossible to conceal my pride, from having in possession an Elegant double gilted mounted small sword – a Trophy of their success.*[13]

Despite the initial surprise of the ambush, the British rapidly advanced and forced Captain Porterfield to withdraw his company to the next American position. Henry Lee wrote that,

> *"The British van pressed forward rapidly and incautiously, until it lined the front of the detachment*

---

[11] Bruce E. Moway, *September 11, 1777: Washington's Defeat at Brandywine Dooms Philadelphia*, (PA: White Mane Books, 2002), 84.
[12] Ibid.
[13] Floyd B. Flickinger, ed., "Heth to Morgan, October 2, 1777, The Diary of Lieutenant William Heth while a Prisoner in Quebec, 1776", *Annual Papers of the Winchester Historical Society*, (1931), 33.

> commanded by Lt. Col. Simms, who poured in a close and destructive fire, and then retreated to the light corps.[14]

Sergeant Thomas Sullivan of Britain's 49th Regiment described another encounter with the American light corps.

> *The Queen's Rangers and Rifle Corps..."* wrote Sullivan, *"advancing to the foot of a hill, saw the enemy formed behind the fence, [and] were deceived by the Rebel's telling them, that they would deliver up their arms; but upon advancing they fired a volley upon our men, and took to their heels, killed and wounded about thirty of the Corps.*[15]

Although this ambush staggered Knyphausen's advance corps, they pressed forward, meeting steady resistance all the way.

> *The enemys Light infantry and Riflemen,"* noted Sergeant Sullivan, *"kept up a running fire, mixed with regular vollies, for 5 miles.*[16]

Major Carl Baurmeister, who was with General Knyphausen, confirmed the duration of the fighting. *"The skirmishing continued to the last hills of Chadd's Ford,"* wrote Baurmeister, after the battle.[17]

---

[14] Lee, 89.
[15] Sullivan, 413.
[16] Ibid.
[17] Baurmeister, 405.

As Knyphausen's advance guard neared the Brandywine, they descended a long hill, and approached a portion of the road that passed through marshy land. Woods and hills bordered the road, providing plenty of cover for Maxwell's troops.[18] *"Heretofore the enemy had been repulsed by our vanguard alone,"* wrote Baurmeister, *"but now the engagement became more serious...."*[19]

General Knyphausen sent a brigade forward to re-enforce his depleted advance corps. Artillery was also placed on a nearby hill and commenced firing at the Americans in the woods and behind some hastily built breastworks on the opposite hill.

> *"We played upon them with two 6 pounders for half an hour,"* recalled Sergeant Sullivan, *"and drove them out of the breastworks, which was made of loose wood upon the declivity of the hill.*[20]

Major Baurmeister gave a more detailed account of this part of the battle. While the artillery bombarded Maxwell's men,

> *The Queen's Rangers...proceeded to the left and after a short but very rapid musketry-fire, supported by the 23rd English Regiment...drove the rebels out of their woods and straight across the lowland. Under cover of a continuous cannonade, the 28th English Regiment went off to the right of the column, and soon the rebels, who had been shouting "Hurrah" and firing briskly from a gorge in front of us, were quickly put to flight.*[21]

---

[18] Ibid.
[19] Ibid.
[20] Sullivan, 413.
[21] Baurmeister, 405.

Sergeant Sullivan, positioned near the center of the attack, recounted the final push that forced Captain Porterfield and the rest of Maxwell's light corps across the Brandywine.

> *As we crossed the brook* [Ring Run] *they formed behind another fence at a field's distance, from whence we soon drove 'em, and a Battalion of Hessians, which formed at the left of our Brigade, fell in with them as they retreated...and after a smart pursuit...they* [the Americans] *crossed the Brandywine and took up post on that side; leaving a few men killed and a few more wounded behind.*[22]

It was only 10:30 in the morning when the last of Maxwell's men crossed the Brandywine and rejoined the American army.[23] They were worn out from several hours of intense fighting. Surprisingly, however, according to Lieutenant Colonel Heth, the losses were, *"inconsiderable in comparison with the Enemys"*.[24] General Washington's aide, Lieutenant Colonel Robert Harrison, concurred with Heth. In a letter to Congress, written that same day, Harrison summed up the morning's action.

> *When I had the Honor of addressing you this morning, I mentioned that the Enemy were advancing and had began a Canonade; I would now beg leave to inform you, that they have kept up a brisk fire from their*

---

[22] Sullivan, 413-414.
[23] Baurmeister, 406.
[24] Flickinger, ed., "William Heth to Col. Daniel Morgan, September 30, 1777, *"The Diary of Lieutenant William Heth while a Prisoner in Quebec, 1777," Annual Papers of the Winchester Historical Society*, 31.

Artillery ever since. Their advanced party was attacked by our light Troops under Genl Maxwell, who crossed the Brandywine for that purpose and had posted his Men on some high Grounds on each side of the Road. The fire from our people was not of long duration as the Enemy pressed on in force, but was very severe. What loss the Enemy sustained cannot be ascertained with precision, but from our situation and briskness of the Attack, it is the general opinion, particularly of those, who were engaged, that they had at least Three Hundred Men killed & wounded. Our damage is not exactly known, but from the best Accounts we have been able to obtain, It does not exceed fifty in the whole.[25]

While it is probable that Harrison's estimate of enemy losses was exaggerated, it is also likely, given the nature of the fight, that Knyphayusen's men sustained much higher casualties than the Americans.

For the next six hours each side remained relatively still. Knyphausen's detachment waited for General Howe's signal to attack, and the Americans waited for Knyphausen to force a passage across the Brandywine. During this pause, General Knyphausen took measures to convince Washington that such an attack was indeed imminent.

The column under Lieut. General Knyphausen," wrote Sergeant Sullivan of the 49th Regiment, *"as had been*

---

[25] Chase, and Lengel, eds., "Lieutenant Colonel Robert Harrison to Congress, September 11, 1777," *The Papers of George Washington, Revolutionary War Series,* Vol. 11, 199.

> *previously conserted, kept the enemy amused in the course of the day, with cannon, and the appearance of forcing the Ford, without intending to pass it, until the attack upon the enemy's right should take place.*[26]

Lieutenant John Marshall, who served in Maxwell's Corps, witnessed Knyphausen's demonstrations.

> *Knyphausen...paraded on the heights, reconnoitred the American army, and appeared to be making dispositions to force the passage of the river.*[27]

Marshall recounted that during this lull, small parties of Americans re-crossed the creek. Scattered firing occurred all day but to little effect. One such incident, however, involving a party of men under Captains Porterfield and Waggoner, proved costly to the British. According to Marshall, who witnessed the incident, Captains Porterfield and Waggoner led a detachment across the creek that,

> *Engaged the British flank guard very closely, killed a captain with ten or fifteen privates, drove them from the wood, and were on the point of taking a field piece. The sharpness of the skirmish soon drew a large body of the British to that quarter, and the Americans were again driven over the Brandywine.*[28]

---

[26] Sullivan, 416.
[27] John Marshall, *The Life of George Washington,* Vol 2, (New York: William Wise & Co., 1925, originally published in 1838), 299.
[28] Ibid, 300.

Washington's aide, Lieutenant Colonel Harrison, may have described the same encounter in his letter to Congress.

> *After the [morning] Affair, the Enemy halted upon the Heights, where they have remained ever since...There has been a scattering loose fire between our parties on each side of the Creek since the Action in the Morning, which just now became warm when Genl Maxwell pushed over with his Corps, and drove them from their Ground with the loss of thirty Men left dead on the Spot, among 'em a Captn of the 49th, and a number of Intrenching Tools with which they were throwing up a Battery.*[29]

It appears that Sergeant Thomas Sullivan also described the encounter, albeit with a much different interpretation of the action. He recorded in his diary that,

> *A company of the 28th and a company of our Regiment advanced upon the hill to the right of the Ford, and in front of the enemy's left flank, in order to divert them, who were posted at a hundred yards distance in their front, behind trees, to the amount of 500, all chosen marksmen. A smart fire maintained on both sides for two hours, without either parties quitting their posts. Out of the two companies there were about 20 men killed and wounded...and two 6 pounders were ordered up the hill to dislodge the enemy if possible...These guns played upon them for some time,*

---

[29] Chase, and Lengel, eds., "Lt. Col. Harrison to Congress, September 11, 1777," *The Papers of George Washington, Revolutionary War Series*, Vol. 11, 199.

> but they were so concealed under cover of the trees, that it was to no purpose...The guns were ordered back and also the two companies in order to draw the enemy after them from the trees, which scheme had the desired effect, for they quitted their post and advanced to the top of the hill where they were attacked [by] four companies of the 10th Battalion, in front, while the 40th made a charge upon their left flank, by going round the hill, and put them to an immediate rout.[30]

It is difficult to determine, with certainty, whether the three accounts are of the same incident. Nonetheless, it is clear that skirmishing continued near Chad's Ford well into the afternoon, and Virginians with Maxwell's Light Corps were heavily involved.

At General Washington's headquarters, conflicting reports of enemy troop movements towards his right flank had Washington concerned. When the reports first arrived Washington saw an opportunity to strike his divided enemy. He ordered Generals Sullivan and Greene to lead their divisions across the creek and attack. But then new reports arrived that denied any such flanking movement. If these reports were true, Sullivan and Greene risked confronting the whole British army. The situation was just too uncertain, so Washington cancelled the attack.

By the early afternoon, however, it was clear that the British were indeed moving against the American right. Believing that it was too late to take the offensive, Washington ordered General Stephen and General Stirling to rapidly march their

---

[30] Sullivan, 414.

divisions northward to the heights near Birmingham Meeting House in order to head off the British. General Sullivan was ordered to follow with his division and assume overall command of the redeployed right wing of Washington's army.

The march route from Brandywine Creek to the heights near Birmingham Meeting House, with its narrow, winding roads over steep, wooded hills and deep ravines tested the stamina of Washington's troops. They doggedly pushed on for three miles to the village of Dilworth, then swung north towards a hill overlooking Birmingham Meeting House and a road running north to Osborne Hill, where General Howe and the British army had paused to rest after their long march around the American army. General Stephen deployed his troops upon a large, cleared, hill just west of the Birmingham Road. General Scott's brigade formed on the left of Woodford's brigade, and General Stirling's division formed to Scott's left.

General Woodford's Virginia brigade thus held the right flank of the American line, which meant he had no support on his right side. To alleviate his concern for his exposed right flank, General Woodford ordered his most experienced regiment, Colonel Thomas Marshall's 3rd Virginia, only 170 strong, to post themselves in an orchard and among the outbuildings of a farm just north of Birmingham Meeting House, several hundred yards in advance and to the right of the American line. From that position they could detect, and possibly stop, any effort of the enemy to get around Woodford's right flank.

The arrival of General Sullivan's division on the left of the American line, however, caused General Stephen and General Stirling to shift their divisions several hundred yards to the right. General Weedon, who did not witness the engagement at Birmingham Heights but likely discussed it with officers who participated in it, explained the impact this shift to the east had on General Woodford's deployment.

> *In making this Alteration, unfavorable Ground, made it necessary for Woodford to move his Brigade 200 Paces back of the Line & threw Marshall's Wood in his front.*[31]

The troops of the 3rd Virginia were no longer on the right of General Woodford's line, they were to his front, exposed but protected by an orchard and stone wall. A mile to their front were approximately 8,000 British and German troops, eager to attack after their day long march, while several hundred yards behind the 3rd Virginians waited about half that number of American troops, determined to offer spirited resistance.[32]

General Howe ordered his advance troops of German jaegers (riflemen) and British light infantry forward from Osborne's Hill shortly after 3:00 p.m. Howe's main body followed about 45 minutes later, marching down Osborne's Hill in several columns before deploying into battle lines as they drew closer to the American line.

In the orchard and farm adjacent to the Birmingham Meeting House, Colonel Marshall's 3rd Virginians waited anxiously for the enemy to come into effective range (approximately 100 yards for muskets, 200 yards for rifles). Captain Johann Ewald, in command of a company of jaegers, led his men straight at the Virginians and described the initial contact.

---

[31] Bob McDonald, transcribed, "Brigadier General George Weedon's Correspondence Account of the Battle of Brandywine, September 11, 1777," Manuscript is held by the Chicago Historical Society.

[32] Thomas McGuire, *The Philadelphia Campaign: Brandywine and the Fall of Philadelphia,* Vol. 1, (Stackpole Books, 2006), 199.

About half past three I caught sight of some infantry and horsemen behind a village on a hill in the distance. I drew up at once and deployed...I reached the first houses of the village with the flankers of the jagers, and Lt. Hagen followed me with the horsemen. But unfortunately for us, the time this took favored the enemy and I received extremely heavy small-arms fire from the gardens and houses, through which, however, only two jagers were wounded. Everyone ran back and I formed them behind the fences and walls at a distance of two hundred paces from the village.[33]

The intensity of the Virginians resistance surprised the British advance troops and they withdrew for the cover of a fence and embankment along the road where they waited for reinforcements. Colonel Marshall's Virginians held firm and waited as well.

Captain Ewald was eventually joined by hundreds of British light infantry (as well as the rest of Howe's main body to his right) whose arrival triggered a withdrawal of the 3rd Virginians from the orchard to a stone wall at the Birmingham Meeting House. Shielded by this wall, the battle tested Virginians maintained such a heavy fire on the enemy that the British troops advancing upon them balked at a frontal assault and maneuvered around their flanks instead.

Praise for the conduct and bravery of the 3rd Virginians at the Birmingham Meeting House was universal. General Weedon noted that Colonel Marshall and his men,

---

[33] Ewald, 84-85.

> *Received the Enemy with a Firmness which will do Honor to him & his little Corps, as long as the 11th of Sepr. is remembered. He continued there ¾ of one Hour, & must have done amazing execution.*[34]

Henry "Light Horse Harry" Lee wrote years later in his memoirs of the war that the 3rd Virginia,

> *Bravely sustained itself against superior numbers, never yielding one inch of ground and expending thirty rounds a man, in forty-five minutes.*[35]

Such resistance though, came at a heavy cost. Over a third of the regiment was lost in the fight, killed, wounded or captured, and the survivors barely escaped.[36] With enemy troops closing in on both sides of the Meeting House wall, the remnants of the 3rd Virginia, led by Colonel Marshall on foot (his horse having been shot out from under him) scurried southward to re-join General Woodford's brigade and continue the fight.

Woodford's troops, with General Scott's Virginians posted to their immediate left, were deployed in a strong position a bit south of the rest of the American line (Stirling's and Sullivan's divisions). A British light infantry officer observed that, "*The position the enemy had taken was very strong indeed -- very commanding ground, a wood on their rear and flanks, a ravine and strong paling* [fence] *in front.*"[37] This British officer was

---

[34] McDonald, transcribed, "Brigadier General George Weedon's Correspondence Account of the Battle of Brandywine, September 11, 1777".
[35] Lee, *The Revolutionary War Memoirs of General Henry Lee*, 89-90.
[36] McGuire, 215.
[37] McGuire, 216.

likely describing the position of General Woodford's troops on the far right of the entire American line. When General Stephen shifted his division right to accommodate the arrival of General Sullivan's division, low ground forced Stephen to deploy his troops 200 yards further south, along higher ground and woods. This turned out to be fortunate for the Americans as the difficult terrain to their front slowed the British advance and helped Woodford and Scott hold firm. Captain John Montresor, General Howe's chief engineer, offered his own description of the terrain defended by Woodford and Scott's troops.

> *The ground on the left being the most difficult the rebels disputed it with the Light Infantry with great spirit, particularly their officers, this spot was a ploughed hill; and they, covered by its summit and flanked by a wood*; *however unfavorable the circumstances* [the ardour of the British lights] *was such that they pushed in upon* [the rebels] *under a very heavy fire.*[38]

It would take more than a direct frontal assault to pry Stephen's division from its strong position. Unfortunately for the Americans, such was not the case on the opposite side of their line, where General Sullivan's division was unable to get into position before the British attacked. While some of Sullivan's men fought well, many fell into disorder at the arrival of the British and withdrew to the rear. General Sullivan, who had taken position near a battery of five cannon in the center of the entire American line to superintend the fight, sent his aides

---

[38] Montresor, 450.

to reform his troops, but they had little success and most of Sullivan's troops withdrew in disarray.

With British troops to their front and now upon their exposed left flank, the pressure on General Stirling's men in the center of the American line increased and they also began to give way. To their right though, remained General Stephen's Virginians with Generals Woodford and Scott at the head of their brigades.

The fight between Woodford's and Scott's Virginians and the British light infantry was severe, made more so upon the British by two of Woodford's cannon that played very effectively upon the British. A British officer who experienced the cannon fire recalled that,

> *There was a most infernal Fire of Cannon & musketry – smoak – incessant shouting – incline to the right! Incline to the Left! – halt! – charge!...The trees* [were] *cracking over ones head. The branches riven by the artillery, the leaves falling as in autumn by the grapeshot.*[39]

Woodford's Virginians were well served by their cannon which, along with heavy musket and rifle fire, initially kept the British and German troops to their front at bay. A German officer noted that, *"The small arms fire was terrible, the counter-fire from the enemy, especially against us, was the most concentrated."*[40]

The collapse of the American left flank and center, however, left Stephen's Virginians in an impossible situation. A British officer described the final assault on Woodford's position.

---
[39] Ibid.
[40] McGuire, 237.

> *They stood the charge till we came to the last* [fence]. *Their line then began to break, and a general retreat took place soon after, except for their guns, many of which were defended to the last, indeed several officers were cut down at the guns.*[41]

General Washington sent help to the right flank in the form of General Greene's division-. It is unclear if Greene's entire division withdrew from the Brandywine to reinforce the crumbling American right flack, or whether it was just General George Weedon's brigade. What is clear though, is that Weedon's Virginians fought well when they arrived in the vicinity of Dilworth to stem the British pursuit of the collapsed American right wing. General Greene recalled that,

> *I marched one brigade of my division, being upon the left wing, between three and four miles in forty-five minutes. When I came upon the ground I found the whole of the troops routed and retreating precipitately, and in the most broken and confused manner. I was ordered to cover the retreat, which I effected in such a manner as to save hundreds of our people from falling into the enemy's hands. Almost all of the park of artillery had an opportunity to get off, which must have fallen into their hands; and the left wing posted at Chads ford, got off by the seasonable check I gave the enemy. We were engaged an hour and a quarter, and lost upwards of a hundred men*

---

[41] Ibid., 238.

killed and wounded. I maintained the ground until dark, and then drew the troops off in good order.[42]

General Weedon, who led the effort to cover the American retreat, humbly provided a brief account of the fight, noting that, *"About 6 General Green's Division arrived to cover the Retreat, one of his brigades (Weedon's) gave the Enemy such a check as produced the desired effect."*[43]

The wording of Weedon's account suggests that General Muhlenberg's brigade, as part of Greene's division, also marched to the relief of the American right wing. Whether they did or not though, it appears that Weedon's brigade did the bulk of the fighting to temporarily hold the British at bay while the American army retreated eastward.

Most of the Virginian troops at the battle of Brandywine, particularly the 3rd Virginia Regiment, fought heroically and suffered heavy casualties for their effort. General Woodford himself was included among the casualties, suffering a gunshot wound to his left hand that threatened to leave him permanently disabled.[44]

General Washington withdrew his battered army to Chester and ordered his brigadier-generals to send as many officers as they thought necessary back along the routes leading to the battle to gather stragglers and return them to the army. *"In*

---

[42] Richard Showman, ed., "General Greene to Henry Marchant, July 25, 1778," *The Papers of General Nathanael Greene,* Vol. 2, (Chapel Hill: University of North Carolina Press, 1980), 471.

[43] McDonald, ed., "Brigadier General George Weedon's Correspondence Account of the Battle of Brandywine, September 11, 1777," Chicago Historical Society. Unpublished.

[44] Mayes, ed., "Edmund Pendleton to William Woodford, November 8, 1777," *The Letters and Papers of Edmund Pendleton,* Vol. 1, 234.

doing this," instructed Washington, *"they will proceed as far, towards the enemy, as shall be consistent with their own safety, and examine every house."*[45] The rest of the American army marched on to Darby and crossed the Schuylkill River via a pontoon bridge, then marched northwest past Philadelphia to the falls of the Schuylkill River near Germanton.

The following day General Washington praised the army in his general orders assuring the troops that they had inflicted far more casualties on the enemy than they had suffered and that the next engagement would prove victorious. Rum was liberally distributed (one gill per man per day while it lasted) and preparations to march the next morning were undertaken.[46]

Washington and his army re-crossed the Schuylkill River the next day and marched southwestward, *"intent on giving the Enemy Battle wherever I should meet them."*[47] General Howe remained in the vicinity of Dilworth for several days, but just as General Washington expected, Howe proceeded to march in a northeasterly direction towards the upper fords of the Schuylkill River.

The two armies nearly clashed a few miles southwest of Valley Forge, but, *"a most violent Flood of Rain,"* damaged the bulk of the American army's musket cartridges and caused Washington to disengage and withdraw northward to Yellow

---

[45] Chase and Lengel, eds., "General Orders, September 12, 1777," *The Papers of George Washington, Revolutionary War Series*, Vol. 11, 204.

[46] Chase and Lengel, eds., "General Orders, September 13, 1777," *The Papers of George Washington, Revolutionary War Series*, Vol. 11, 212.

[47] Chase and Lengel, eds., "General Washington to John Hancock, September 23, 1777," *The Papers of George Washington, Revolutionary War Series*, Vol. 11, 301.

Springs.[48] Concerned that General Howe was maneuvering to flank him (as he had at Brandywine and Brooklyn) Washington marched further northwestward.[49] Reports that the enemy was actually marching towards Swedes Ford on the Schuylkill River, however, just a few miles from Valley Forge, prompted Washington to rush eastward to re-cross the river and head off General Howe. The Americans arrived at Fatland Ford, just north of Valley Forge, on September 20th, and deployed for several miles up and down the Schuylkill River to guard against a British crossing.[50]

Philadelphia Falls

General Washington was unsure of General Howe's true objective; the British commander had placed his army about mid-way between the American capital at Philadelphia and a vital American supply depot at Reading. Having been burned twice by General Howe's flanking movements, Washington was determined not to be outflanked again. When British troop movements on September 21, suggested that Reading was Howe's true objective, Washington acted quickly and shifted his army towards Reading.[51]

Unfortunately for the Americans, General Washington had miscalculated. General Howe was not interested in the supplies at Reading; Philadelphia was his true objective, and his army reversed direction and crossed the Schuylkill River at Fatland

---

[48] Ibid.
[49] Chase and Lengel, eds., "General Washington to John Hancock, September 18, 1777," *The Papers of George Washington, Revolutionary War Series*, Vol. 11, 262.
[50] McGuire, 306.
[51] Ibid, 320.

Ford unopposed.[52] General Washington sheepishly informed Congress, which had removed to Lancaster days earlier, of this development.

> *The Enemy, by a variety of perplexing Maneuvres thro' a County from which I could not derive the least intelligence being to a man disaffected, contrived to pass the Schuylkill last Night at the Flat land and other Fords in the Neighbourhood of it. They marched immediately towards Philada and I imagine their advanced parties will be near that City to Night. They had so far got the Start before I recd certain intelligence...that I found it in vain to think of overtaking their Rear with Troops harassed as ours had been with constant marching since the Battle of Brandywine....*[53]

General Washington felt obligated to offer an explanation of how he found himself out of position to challenge the British crossing of the Schuylkill.

> *The day before yesterday they were again in motion and marched rapidly up the Road leading towards Reading. This induced me to believe that they had two objects in view, one to get round the right of the Army, the other perhaps to detach parties to Reading where we had considerable quantities of military Stores. To frustrate those intentions I moved the Army up on this side of the River to this place* [Pottsgrove] *determined to keep pace*

---

[52] Ibid, 322.
[53] Chase and Lengel, eds., "General Washington to John Hancock, September 23, 1777," *The Papers of George Washington, Revolutionary War Series*, Vol. 11, 301-302.

> with them, but early this morning I recd intelligence that they had crossed at the Fords below.[54]

Addressing a question Washington was sure the entire Congress was thinking, he added,

> *Why I did not follow immediately I have mentioned in the former part of my letter. But the strongest Reason against being able to make a forced march is the want of Shoes;...at least one thousand Men are bare footed and have performed the marches in that condition.*[55]

General Washington was not alone in his assessment that the American army was in no condition to undertake a forced march to catch and fight General Howe's army. A War Council of his general officers unanimously agreed that, *"from the present state of the Army, it would not be adviseable to advance upon the Enemy...till the detachments and expected Reinforcements come up."*[56] Five days later, at a second war council, Washington's commanders proposed that the army move closer to the British in Philadelphia and await either reinforcements or a more favorable opportunity to strike.[57] While they waited, news from the north gave them hope of a much needed victory over the British.

---

[54] Ibid.
[55] Ibid.
[56] Chase and Lengel, eds., "Council of War, September 23, 1777," *The Papers of George Washington, Revolutionary War Series*, Vol. 11, 295-296.
[57] Chase and Lengel, eds., "Council of War," September 28, 1777," *The Papers of George Washington, Revolutionary War Series*, Vol. 11, 338-339.

Burgoyne's March Route 1777

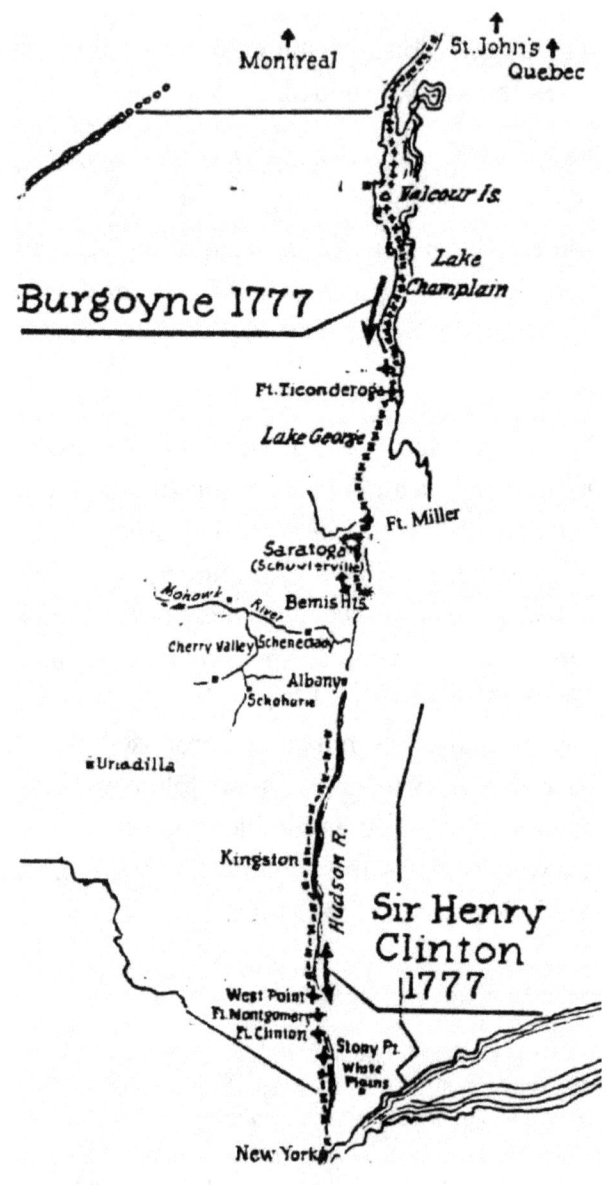

Chapter Nine

Battle of Saratoga

For much of the summer of 1777, General Washington had received disturbing reports from General Philip Schuyler in New York about a large British army under General John Burgoyne advancing unhindered down Lake Champlain and the Hudson River Valley from Canada. Burgoyne's objective was Albany, from which he hoped to sever New England from the rest of the colonies with an army of 7,000 men, aided by one thousand Indians who spread terror along their march.

Schuyler informed General Washington in early August that he was powerless to halt Burgoyne's advance through New York in part because fear of the Indians had infected his troops:

> *The most unaccountable panic has seized the Troops…A few shot from a small party of Indians has more than once thrown them into the greatest Confusion – The Day before Yesterday three hundred of our Men…came running in, being drove by a few Indians, certainly not more than fifty.*[1]

Schuyler also complained that he lacked enough troops to stop Burgoyne.[2]

---

[1] Grizzard, Jr., ed., "General Philip Schuyler to General George Washington, August 1, 1777," *The Papers of George Washington, Revolutionary War Series,* Vol. 10, 482-483.

[2] Grizzard, Jr., ed., "General Philip Schuyler to General George Washington, August 13, 1777," *The Papers of George Washington, Revolutionary War Series*, Vol. 10, 606.

General Washington responded to Schuyler's appeals in mid-August and sent reinforcements, including Colonel Morgan's Rifle Corps, to New York specifically to counteract the Indians. Washington expressed his high regard of Morgan and his men in his orders to Colonel Morgan.

> *You will march...with* [your] *corps to Peekskill, taking with you all the baggage belonging to it. When you arrive there, you will take directions from General Putnam, who, I expect, will have vessels provided to carry you to Albany. The approach of the enemy in that quarter has made a further reinforcement necessary. I know of no corps so likely to check their progress, in proportion to its number, as that under your command. I have great dependence on you, your officers and men, and I am persuaded you will do honor to yourselves, and essential services to your country.*[3]

General Washington also expressed his confidence in Morgan's Rifle Corps to New York Governor George Clinton:

> *I am forwarding as fast as possible, to join the Northern army, Colonel Morgan's corps of riflemen, amounting to about five hundred. These are all chosen men, selected from the army at large, well acquainted with the use of rifles, and with that mode of fighting, which is necessary to make them a good counterpoise to the Indians; and they have distinguished themselves on a variety of occasions,*

---

[3] Grizzard, Jr., ed., "General Washington to Colonel Daniel Morgan, August 16, 1777," *The Papers of George Washington, Revolutionary War Series,* Vol. 10, 641.

> since the formation of the corps, in skirmishes with the enemy. I expect the most eminent services from them, and I shall be mistaken if their presence does not go far towards producing a general desertion among the savages.[4]

Colonel Morgan's orders were slightly adjusted during his march north to New York. Congress removed General Schuyler from command of the northern army in mid-August and installed General Horatio Gates in his place. Morgan was instructed to report directly to General Gates, a fellow resident of the Shenandoah Valley and former officer in the British army who had settled in Virginia in 1773 after leaving the army.

General Gates, who was encamped with the northern army a few miles north of Albany, was very pleased to hear of the rifle corps' transfer and expressed his satisfaction in a letter to General Washington.

> I cannot sufficiently thank your Excellency for sending Colonel Morgan's corps to this army; they will be of the greatest service to it, for until the late successes this way, I am told the army were quite panic-struck by their Indians, and their Tory and Canadian assassins in Indian dresses.[5]

The "late successes" that Gates referred to included the defeat of a British detachment near Fort Stanwix along the Mohawk River (that was cooperating with General Burgoyne's invasion) and the stunning American victory near Bennington,

---

[4] Grizzard, Jr., ed., "General Washington to George Clinton, August 16, 1777," *The Papers of George Washington, Revolutionary War Series*, Vol. 10, 636.

[5] Philander D. Chase and Edward G. Lengel, eds., "General Gates to General Washington, August 22, 1777," *The Papers of George Washington, Revolutionary War Series*, Vol. 11, 38.

Vermont where 2,000 New Hampshire and Vermont militia overwhelmed a 900 man foraging party from Burgoyne's army.[6] These victories significantly improved American morale in late August, which in turn led to an increase in militia turnout for General Gates. By the time Morgan and his riflemen joined Gates in early September, they found an American army that had surpassed Burgoyne's in size.[7]

Unfortunately, Morgan's corps was not at full strength when it reached camp. Three months of active service and the long journey north took a toll on the rifle corps. Less than 400 riflemen arrived with Colonel Morgan fit for service.[8] General Gates partially alleviated Morgan's manpower shortage by drafting fifteen of the most hardy musket-men from each regiment in his army to serve in a corps of light infantry.[9] They were led by Major Henry Dearborn of New Hampshire, a veteran of Arnold's expedition to Quebec and an officer whom Morgan respected. The addition of 250 hand-picked musket-men with bayonets greatly enhanced the fighting effectiveness of Morgan's riflemen. Morgan's combined corps had both long range and close combat capabilities and would soon have the opportunity to demonstrate these capabilities on the battlefields of Saratoga.

---

[6] Boatner, III, ed., *Encyclopedia of the American Revolution, 3rd ed.*, 75.

[7] Lesser, ed., "A General Return of the Continental Troops Under the Command of Major General Horatio Gates, September 7, 1777," (Gates Papers) *The Sinews of Independence: Monthly Strength Reports of the Continental Army*, 49.

[8] General James Wilkinson, "A Return of Colonel Morgan's detachment of Riflemen, September 3, 1777," *Memoirs of My Own Times,* Vol. 1 Appendix C.

[9] James Graham, "General Gates to Colonel Morgan, August 29, 1777," *The Life of General Daniel Morgan*, (Bloomingburg, NY: Zebrowski Historical Services Publishing Company, 1993), 138.

With the influx of reinforcements swelling his ranks, General Gates advanced north with his army, halting on September 12th, upon Bemis Heights, an excellent defensive position overlooking the Hudson River. Most of the American troops spent the next week erecting fortifications. Colonel Morgan's corps had a different task, reconnaissance. With General Burgoyne and his army just a few miles to the north preparing for a final push on Albany, General Gates wanted as much intelligence on them as possible. Frequent patrols by Morgan's troops helped provide that intelligence.

General Burgoyne also desired intelligence on the Americans, but the departure of most of his Indians in late August hampered his ability to collect it. Dissatisfaction with the campaign and their treatment by Burgoyne caused most of natives to leave, and with them went Burgoyne's best scouts. The British general knew that a large enemy force lay to his front somewhere north of Albany, but he was unsure of its strength and placement. Undeterred by the uncertainty of what awaited him in the woods to his south, Burgoyne boldly divided his army into three columns and advanced towards the Americans on September 19th.[10]

General Burgoyne's left column, commanded by General von Riesdel, comprised approximately 3,000 men.[11] It included most of the artillery and a large baggage train protected by four German regiments and the 47th British regiment. This column slowly marched along the river road towards the Americans at Bemis Heights, stopping several

---

[10] James Baxter, ed., *The British Invasion from the North: Digby's Journal of the Campaigns of Generals Carleton and Burgoyne from Canada, 1776-1777*, (New York: De Capa Press, 1970), 267.

[11] John Luzader, *Saratoga: A Military History of the Decisive Campaign of the American Revolution*, (New York: Savas Beatie, 2008), 230.

times to repair bridges destroyed by the Americans. General Burgoyne's right column, approximately 2,550 men strong under General Simon Fraser, was tasked to screen the British right flank and probe the American left flank.[12] To do this, Fraser marched nearly three miles west, away from the river, and then swung south towards the Americans.

General Burgoyne marched southwestward into the woods with his center column, moving in a diagonal direction from the river. This column totaled 1,600 men from the 9th, 20th, 21st, and 62nd British regiments and was commanded by General James Hamilton. Four cannon joined the column.[13]

The three columns began their march around 9:00 a.m. on September 19, and were quickly observed by American scouts. Word reached General Gates in the American camp that the enemy was on the move. Colonel James Wilkinson, an aide to General Gates, recalled that General Gates,

> *Ordered Colonel Morgan to advance his corps, who was instructed, should he find the enemy approaching, to hang on their front and flanks, to retard their march, and cripple them as much as possible.* [14]

Morgan's light corps, numbering around 600 men, advanced in a narrow column through the thick woods towards Burgoyne's center detachment, the riflemen ahead of the light infantry.[15] They marched about a mile and a half and emerged

---

[12] Ibid.
[13] Richard M. Ketchum, *Saratoga,: Turning Point of America's Revolutionary War*, (NY: Holt & Co., 1997), 357.
[14] Wilkinson, 236.
[15] Wilkinson, Appendix E and Joseph Lee Boyle, ed., "From Saratoga to Valley Forge: The Diary of Lt. Samuel Armstrong," *The Pennsylvania Magazine of History and Biography,* Vol. 121, No. 3 (July 1997), 245.

onto the edge of an abandoned farm. The clearing was dotted with tall grass, dead trees and stumps. Two small buildings, described as cabins by eyewitnesses, sat on a rise of ground about 300 yards away. The opposite wood line was only 150 yards beyond the cabins. Morgan's corps arrived at the southern end of the clearing just as an enemy advance party attacked the American picquet guard posted in the cabins. Samuel Armstrong, a member of Major Dearborn's light infantry, described the encounter:

> [At] *about 12 Oclock we were Alarm'd by the firing of two or three Musketts from the Enemies Scouts, upon which the Riffle and Light Infantry Battalions were Ordered off to Scour the Woods. We forwarded down to our Picquet Guard where we had no sooner got Sight of than we saw the Enemy surrounding them.* [16]

The American pickets in the cabins quickly dispersed and fled Freeman's Farm in the face of the approaching enemy.[17] As the British skirmishers pushed past the cabins and approached Morgan's position in the southern wood line they collided with the advance of Morgan's light corps. British Lieutenant William Digby described what happened:

> *A little after 12 our advanced picquets came up with Colonel Morgan and engaged, but from the great superiority of fire received from him – his numbers being*

(Henceforth referred to as Lieutenant Armstrong's Diary)
[16] Boyle, ed., "Lieutenant Armstrong's Diary," 245.
[17] John Burgoyne, *A State of the Expedition from Canada*, (New York Times & Arno Press, 1969), 68.

much greater – they were obliged to fall back, every officer being either killed or wounded except one.[18]

The British skirmishers, outnumbered and outgunned, retreated under a deadly barrage of fire from Morgan's men. Advance elements of Morgan's corps pursued the fleeing skirmishers across the field and into the woods beyond. Their pursuit abruptly ended when the Americans discovered the main body of General Burgoyne's center column. These soldiers, deployed in the opposite wood line from which Morgan's men emerged, startled the Americans with a volley that hit friend and foe alike.[19]

The result was chaos for Morgan's corps. Already somewhat disorganized by the unauthorized charge across the open field, Morgan's corps disintegrated on contact with the main body of the enemy. Men ran in all directions to escape. The sudden emergence on their left flank of two British companies and a field piece from General Fraser's column added urgency to the flight of Morgan's men.[20]

Appalled by the turn of events, Colonel Morgan struggled to reorganize his shattered corps. He used an uncommon military tool to do so, a turkey whistle. Colonel Wilkinson appeared on the scene at this time and observed Morgan's efforts to reform his riflemen:

---

[18] Baxter, ed., "Digby's Journal," 272.
[19] Horatio Rogers ed., *Hadden's Journal and Orderly Book: A Journal Kept in Canada and Upon Burgoyne's Campaign in 1776 and 1777*, (Boston: Gregg Press, 1972), 163.
[20] Sydney Jackman, ed., *With Burgoyne from Quebec: An Account of the Life at Quebec and of the Famous Battle at Saratoga*, (Toronto: Macmillan of Canada, 1963), 72.
   (Henceforth referred to as Anburey's Journal)

> *My ears were saluted by an uncommon noise, which I approached, and perceived Colonel Morgan attended by two men only, who with a turkey call was collecting his dispersed troops. The moment I came up to him, he burst into tears, and exclaimed, 'I am ruined, by G—d! Major Morris ran on so rapidly with the front, that they were beaten before I could get up with the rear, and my men are scattered God knows where.' I remarked to the Colonel that he had a long day before him to retrieve an inauspicious beginning, and informed him where I had seen his field officers, which appeared to cheer him....* [21]

One of the field officers that Colonel Wilkinson met prior to Morgan was Major Joseph Morris. Morris led the charge against the fleeing British pickets and gave Wilkinson a detailed account of the engagement:

> *From him* [Major Morris] *I learnt that the corps was advancing by files in two lines, when they unexpectedly fell upon a picket of the enemy, which they almost instantly forced, and pursuing the fugitives, their front had as unexpectedly fallen in with the British line; that several officers and men had been made prisoners, and that to save himself, he had been obliged to push his horse through the ranks of the enemy, and escaped by a circuitous route.* [22]

Wilkinson also encountered Lieutenant Colonel Richard Butler, the rifle corps' second in command. He confirmed Morris's account.

---

[21] Wilkinson, 238.
[22] Ibid, 237.

I [Wilkinson] *crossed the angle of the field, leapt the fence, and just before me on a ridge discovered Lieutenant-colonel Butler with three men, all tree'd; from him I learnt that they had 'caught a Scotch prize,' that having forced the picket, they had closed with the British line, had been instantly routed, and from the suddenness of the shock and the nature of the ground, were broken and scattered in all directions.*[23]

Fortunately for Colonel Morgan, the British did not pursue his scattered troops and a pause in the engagement allowed Morgan to re-organize most of his riflemen in the woods to the south of Freeman's Farm, their front protected by a deep ravine.[24] To Morgan's left, reinforcements in the form of three continental regiments from New Hampshire from General Enoch Poor's brigade deployed westward. Major Dearborn's light infantry, which had become separated from Morgan and his riflemen in the initial engagement, covered the far left flank of the American line and fought the rest of the day detached from Morgan and his riflemen.

The battle at Freeman's Farm resumed around mid-afternoon when General Hamilton's troops emerged from the far woods and advanced across Freeman's field towards the Americans. The British concentrated on a small knoll just beyond the cabins. Lieutenant James Hadden, a British artillery officer, described the action:

---

[23] Ibid.
[24] Rogers ed., *Hadden's Journal and Orderly Book: A Journal Kept in Canada and Upon Burgoyne's Campaign in 1776 and 1777*, 164.

> *The Enemy being in possession of the wood almost immediately attacked the Corps which took post beyond two log Huts on Freemans Farm...I was advanced with two Guns to the left of the 62nd Regt and ye two left companies being formed en potence [refused or bent to protect the flank] I took post in the Angle...In this situation we sustained a heavy tho intermitting fire for near three hours....*[25]

The American fire, enhanced by the accuracy of Morgan's riflemen (some of who climbed trees to get better shots) was especially hard on the British artillerymen. Lieutenant Hadden lost 19 out of 22 men and all of his horses. The 62nd regiment lost nearly half of its men.[26]

Hadden's position was not the only hot spot for the British. The battle raged all along the line. British Lieutenant William Digby noted that he had never seen anything like it:

> *Such an explosion of fire I never had any idea of before, and the heavy artillery joining in concert like great peals of thunder, assisted by the echoes of the woods, almost deafened us with the noise.*[27]

British corporal Roger Lamb gave a similar account:

> *The conflict was dreadful; for four hours a constant blaze of fire was kept up, and both armies seemed to be determined on death or victory...Men, and particularly officers, dropped every moment on each*

---

[25] Ibid. 165.
[26] Ibid.
[27] Baxter, ed., "Digby's Journal," 237.

side. Several of the Americans placed themselves in high trees, and as often as they could distinguish a British officer's uniform, took him off by deliberately aiming at his person.[28]

The impact of American marksmanship, no doubt enhanced by Colonel Morgan's riflemen, was also noted by Colonel James Wilkinson, who observed that it repeatedly drove the British from the Freeman house hill:

The fire of our marksmen from this wood was too deadly to be withstood by the enemy in line, and when they gave way and broke, our men rushing from their cover, pursued them to the eminence, where having their flanks protected, they [the enemy] rallied and charging in turn drove us back into the wood, from whence a dreadful fire would again force them to fall back; and in this manner did the battle fluctuate, like waves of a stormy sea, with alternate advantage for four hours without one moment's intermission. The British artillery fell into our possession at every charge, but we could neither turn the pieces upon the enemy, nor bring them off...The slaughter of this brigade of artillerists was remarkable, the captain and thirty-six men being killed or wounded out of forty-eight.[29]

---

[28] Roger Lamb, *An Original and Authentic Journal of Occurrences During the Late American War from Its Commencement to 1783*, (Dublin: Wilkinson & Courtney, 1809), 159.
    Reprinted by Arno Press, 1968.
[29] Wilkinson, 241.

Even General Burgoyne acknowledged the impact of Morgan's riflemen from Virginia and Pennsylvanian riflemen.

> *The enemy had with their army great numbers of marksmen, armed with rifle-barrel pieces; these during an engagement, hovered upon the flanks in small detachments, and were very expert in securing themselves, and in shifting their ground. In this action many placed themselves in high trees in the rear of their own line, and there was seldom a minute's interval of smoke, in any part of our line without officers being taken off by single shot. It will naturally be supposed, that the Indians would be of great use against this mode of fighting. The example of those that remained after the great desertion proved the contrary, for not a man of them was to be brought within the sound of a rifle shot.* [30]

As sunset approached, the British were in serious trouble. The 62nd regiment was shattered, and the other regiments were barely holding on. Suddenly, drums were heard in the woods beyond Morgan's right flank. German reinforcements from the river column under General Riedesel emerged from the woods to Morgan's right and onto the field to relieve Hamilton's wavering center column. A German artillery officer, Captain Pausch, recalled

> *I had shells brought up and placed by the side of the cannon and as soon as I got the range, I fired twelve*

---

[30] Burgoyne, 39-40.

or fourteen shots in quick succession into the foe who were within good pistol shot distance.[31]

The targets of the German shelling included Morgan's riflemen, who had been on the scene for over six hours.

The arrival of the German reinforcements revived the spirits of General Hamilton's battered British troops and they rallied one more time. Captain Pausch noted

The firing from muskets was at once renewed, and assumed lively proportions, particularly the platoon fire from the left wing of Riedesel. Presently, the enemy's fire, though very lively at one time, suddenly ceased. I advanced about sixty paces sending a few shells after the flying enemy, and firing from twelve to fifteen shots more into the woods into which they had retreated. Everything then became quiet; and about fifteen minutes afterwards darkness set in.... [32]

One of the most intense battles of the Revolutionary War was over, and the carnage was appalling. The field was littered with dead and wounded men who remained unattended all night. British lieutenant William Digby described the scene:

During the night we remained in our ranks, and tho we heard the groans of our wounded and dying at a small distance, yet could not assist them till morning, not knowing the position of the enemy, and expecting

---

[31] George Pausch, *Journal of Captain Pausch, Chief of the Hanau Artillery During the Burgoyne Campaign*, Translated by William L. Stone, (Albany, NY: Joel Munsell's Sons, 1886), 137-138.

[32] Ibid. 138.

the action would be renewed at day break. Sleep was a stranger to us…
20th. At day break we sent out parties to bring in our wounded, and lit fires as we were almost froze with cold, and our wounded who lived till the morning must have severely felt it. [33]

British ensign Thomas Anburey had the misfortune to command a burial party the next day.

The day after our late engagement, I had as unpleasant a duty as can fall to the lot of an officer, the command of the party sent out to bury the dead and bring in the wounded…They [the wounded] *had remained out all night, and from the loss of blood and want of nourishment, were upon the point of expiring with faintness; some of them begged they might lie and die, others again were insensible, some upon the least movement were put in the most horrid tortures, and all had near a mile to be conveyed to the hospitals; others at their last gasp, who for want of our timely assistance must have inevitably expired. These poor creatures, perishing with cold and weltering in their blood, displayed such a scene, it must be a heart of adamant that could not be affected at it.* [34]

Although the British kept the field, it was at a heavy cost; they suffered twice as many casualties as the Americans. Some

---

[33] Baxter, ed., "Digby's Journal," 274.
[34] Jackman, ed., "Anburey's Journal," 176.

of the British, like Thomas Anburey, questioned the value of the victory:

> *Notwithstanding the glory of the day remains on our side, I am fearful the real advantage resulting from this hard fought battle will rest on that of the Americans, our army being so weakened by this engagement as not to be of sufficient strength to venture forth and improve the victory, which may, in the end, put a stop to our intended expedition; the only apparent benefit gained is that we keep possession of the ground where the engagement began.* [35]

General Burgoyne, in a letter to Lord George Germain, reached a similar conclusion about the victory:

> *It was soon found that no fruits, honour excepted, were attained by the preceding victory, the enemy working with redoubled ardour to strengthen their left, their right was already unattackable.* [36]

Despite their retreat from the field, the attitude in the American camp was far from defeatist. In fact, most American accounts bragged about punishing the enemy and attributed the retreat merely to darkness. Major Henry Dearborn's observation was typical:

> *On this Day has Been fought one of the Greatest Battles that Ever was fought in America, & I Trust we have Convinced the British Butchers that the Cowardly*

---

[35] Ibid, 175.
[36] Burgoyne, "General Burgoyne to Lord Germaine, October 10, 1777," *A State of the Expedition,* Appendix, 88.

yankees Can & when there is a Call for it, will, fight...The Enimy Brought almost their whole force against us, together with 8 pieces of Artillery. But we who had Something more at Stake than fighting for six Pence Pr Day kept our ground til Night Closed the scene, & then Both Parties Retire'd.[37]

Many of the British did indeed change their opinion of the Americans after the battle. Ensign Thomas Anburey's comments were typical:

The courage and obstinacy with which the Americans fought were the astonishment of everyone, and we now become fully convinced they are not that contemptible enemy we had hitherto imagined them, incapable of standing a regular engagement, and that they would only fight behind strong and powerful works.[38]

Colonel Morgan's riflemen were some of the combatants that Anburey referred to. They engaged the British for approximately six hours and inflicted heavy losses on them. They, in turn, suffered only sixteen casualties (seven killed and nine wounded).[39] The extended range of rifles, which allowed Morgan's men to fire from beyond musket distance, contributed to the low rifle casualties. In contrast, Major Dearborn's light infantry battalion, armed with smoothbore muskets, had the highest number of unit deaths, with eighteen. Twenty-two of

---

[37] Lloyd Brown and Howard Peckman, ed., "September 19, 1777," *Revolutionary War Journals of Henry Dearborn, 1775-1783*, (Chicago: The Caxton Club, 1939), 106.
[38] Jackman, ed., "Anburey's Journal," 175.
[39] Wilkinson, Appendix D.

his men were wounded.[40] An official count of American casualties listed 321 in all, with 65 killed, 218 wounded, and 38 missing.[41]

Although the Americans believed they had dealt Burgoyne a significant blow, they realized that his army was still very dangerous and braced themselves for another attack. Fortunately for the Americans -- who were very low on ammunition -- it was delayed by several weeks.

General Burgoyne actually planned to resume his advance the next day, but canceled at the last minute to rest his troops. While they rested, Burgoyne received news that General Henry Clinton was leading a British detachment northward from New York to attack the American posts in the New York Highlands and draw some of the American troops with General Gates away from Bemis Heights. Although Clinton's force was too small to fight its way to General Burgoyne, both generals hoped that Clinton's presence on the Hudson River would force General Gates to send some of his troops south and give Burgoyne a better chance to break through to Albany.

General Burgoyne decided to fortify his position and wait for Clinton's advance to have the desired effect. Unfortunately for Burgoyne, few Americans left Bemis Heights. In fact, during the seventeen day pause, the American army swelled to over 10,000 men.[42]

With time on his side, the ever cautious Gates waited behind his fortified lines. Every passing day saw Burgoyne's supplies dwindle and his situation grow more desperate. Colonel

---

[40] Ibid.
[41] Ibid.
[42] Wilkinson, "A General Return of the Army of the United States, commanded by the Hon. Major-General Horatio Gates, Oct. 4, 1777," Appendix E.

Morgan's light corps added to Burgoyne's discomfort by constantly harassing his lines and foraging parties. General Burgoyne acknowledged Morgan's impact in a letter.

> *From the 20th of September to the 7th of October, the armies were so near, that not a single night passed without firing, and sometimes concerted attacks upon our advanced picquets; no foraging party could be made without great detachments to cover it; it was the plan of the enemy to harass the army by constant alarms, and their superiority of numbers enabled them to attempt it without fatigue to themselves.*[43]

The value of Morgan's riflemen was highlighted in an exchange of letters between General Washington and General Gates. On September 24, General Washington, who had suffered a significant defeat against General Howe at the Battle of Brandywine in Pennsylvania, congratulated Gates on his success at Freeman's Farm. He then requested the return of Morgan's riflemen to the main army:

> *This Army has not been able to oppose General Howe's with the success that was wished, and needs a Reinforcement. I therefore request, if you have been so fortunate, as to Oblige General Burgoyne to retreat to Tyconderoga—or If you have not and circumstances will admit, that you will Order Colo. Morgan to Join me again with his Corps. I sent him up when I thought you materially wanted him, and if his services can be*

---

[43] Burgoyne, 168.

> *dispensed with now, you will direct his return immediately.*[44]

The fact that Washington requested only the Rifle Corps return is a testament of his high regard for the unit. General Gates's response was equally telling of his esteem and reliance on the riflemen.

> *Since the Action of the 19th Instant, the Enemy have kept the Ground they Occupied the Morning of that Day, And fortified their Camp. The Advanced Centrys of my picquets, are posted within Shot, And Opposite the Enemy's; neither side have given Ground an Inch. In this Situation, Your Excellency would not wish me to part with the Corps the Army of General Burgoyne are most Afraid of.*[45]

General Gates added that with British provisions dwindling it was only a matter of days or weeks before Burgoyne either risked another battle or withdrew to Ticonderoga. Gates was confident of success and informed Washington that he hoped to soon send far more than just Morgan's Corps southward to reinforce Washington.[46]

The day after he wrote to Washington, General Gates ordered Colonel Morgan to reconnoiter the enemy's lines with his light corps. They circled around to the rear of the British and captured several prisoners.[47] Major Dearborn, who nearly

---

[44] Chase and Lengel, eds., "General Washington to General Gates, September 24, 1777," *The Papers of George Washington*, Vol. 11, 310.
[45] Chase and Lengel, eds., "General Gates to General Washington, October 5, 1777," *The Papers of George Washington*, Vol. 11, 392.
[46] Ibid.
[47] Brown and Peckman, ed., "September 30, 1777," *Revolutionary War Journals of Henry Dearborn,* 108.

forty years later would gently criticize Morgan in his memoirs as, *"a brave officer in action,* [but] *too cautious as a partisan,"* recorded in his diary on October 6, that they got lost, due to heavy rain and the darkness of the night, on their way back to camp and were forced to spend the night in the woods.[48]

While Morgan, Dearborn, and the light troops endured a miserable night in the field, the critical British supply situation finally forced General Burgoyne to act. Unwilling to accept the defeat of his plans against Albany, Burgoyne decided to advance and probe the American position on Bemis Heights with a large detachment in hopes of discovering a weak spot.

Battle of Bemis Heights

General Burgoyne's reconnaissance force numbered over 1,500 men and ten cannon.[49] Although nearly all of the army's units contributed men, the bulk came from the right wing of Burgoyne's line. Two redoubts anchored this position. One was manned by British light infantry under Lieutenant Colonel Balcarress. The other was defended by German grenadiers under Lieutenant Colonel Breymann. Since the march route of the British detachment placed this force between the Americans and the redoubts, General Burgoyne drew heavily from these fortifications, leaving them lightly manned.[50]

---

[48] Ibid.
[49] Eric Schnitzer, "Battling for the Saratoga Landscape," *Cultural Landscape Report: Saratoga Battle, Saratoga National Park,* Vol. 1 (Boston, MA: Olmsted Center for Landscape Preservation), 44.
[50] Henry Dearborn, "A Narrative of the Saratoga Campaign – Major General Henry Dearborn, 1815," *The Bulletin of the Fort Ticonderoga Museum,* Vol. 1, No. 5, January, 1929, 7, and Brown and Peckman, ed., "September 30, 1777," *Revolutionary War Journals of Henry Dearborn ,*108.

Burgoyne led his troops out of camp around noon and slowly advanced toward the American left wing on Bemis Heights. His skirmishers drove off American picquets less than a mile into their march. Burgoyne halted in a wheat field and posted his men in a long line facing south, towards Bemis Heights. The British right flank, composed of light infantry troops, rested on a ridge just east of a wooded hill. German troops, supported by artillery, held the center of the line, and the left was defended by British grenadiers and artillery.[51]

General Burgoyne tried to observe the American fortifications from the wheat field, but the woods obscured his view. Ironically, as Burgoyne and his staff struggled to peer through the woods, they were observed by an American officer.

When reports of Burgoyne's advance reached American headquarters, General Gates dispatched his aide, Lieutenant Colonel Wilkinson, to investigate. Wilkinson reported that the enemy was on the move, at which General Gates sent Wilkinson to Colonel Morgan with instructions to *"begin the game."* Wilkinson recalled

> *I waited on the Colonel,* [Morgan] *whose corps was formed in front of our centre, and delivered the order; he knew the ground, and inquired the position of the enemy: they were formed across a newly cultivated field, their grenadiers with several pieces on the left, bordering on a wood and a small ravine...their light infantry on the right, covered by a worm fence at the foot of the hill...thickly covered with wood; their centre composed of British and German battalions. Colonel Morgan, with his usual sagacity, proposed to make a circuit with his corps by our left, and*

---

[51] Luzader, 52.

under cover of the wood to gain the height on the right of the enemy, and from thence commence his attack, so soon as our fire should be opened against their left.[52]

According to Major Wilkinson, General Gates approved Morgan's proposal and ordered General Poor's brigade to attack Burgoyne's left flank. General Learned's brigade followed, with instructions to strike the center of the enemy.[53]

Colonel Morgan's corps was still moving into position along a wooded ridge overlooking Burgoyne's right flank when fighting erupted on the British left. It was General Poor's men, followed by General Learned's troops, and the intensity of the engagement caused some in Morgan's corps to worry that their comrades were losing. Major Dearborn recalled,

Our light troops moved on with a quick step in the course directed, and after ascending the woody hill to a small field about 500 yards to the right of the

---

[52] Wilkinson, 268.

[53] Note: Historians have long reported, based largely on the memoirs of Major Wilkinson, that General Benedict Arnold had relinquished command of his division after the Battle of Freeman Farm following a heated dispute with General Gates over the lack of credit Arnold received in General Gates's report to Congress and the removal of Colonel Morgan's light corps from Arnold's command. The long held belief was that Arnold was preparing to leave the army and present his grievances about General Gates to Congress when, upon learning of Burgoyne's advance towards the American lines, Arnold mounted a horse (some claim while intoxicated) and rode out of camp (without the authorization or approval of General Gates) to assume command of his old brigades. The recent discovery of a letter written on October 9, 1777 from Nathaniel Bacheller, an adjutant in a New Hampshire militia battalion attached to Learned's brigade, claims that General Arnold sought and received permission from General Gates to lead troops against Burgoyne.
See: Nathaniel Bacheller Letter, October 9, 1777, Copy on file at Saratoga National Historical Park.

> Enemies main line, we discovered a body of British light Infantry handsomely posted on a ridge 150 yards from the edge of the wood where we then were. At this time the fire of the two main armies was unusually heavy and we were apprehensive from the fire that our line was giving way.[54]

Colonel Morgan rushed his men towards the enemy flank. Captain Thomas Posey of Virginia described what happened:

> They [the enemy] had repulsed [General] Arnold twice before Morgan made his attack, which was on the right wing of [the] enemy – the [rifle] regiment had march'd under cover of a thick wood, and a ridge, which ridge the enemy were about to take possession of as Morgan gained the summit of it, the enemy being within good rifle shot, the regiment poured in a well directed fire which brought almost every officer on horseback to the ground.[55]

Lieutenant Colonel Richard Butler noted the impact of Morgan's attack on the battle.

> I had the Honour to lead the Corps of Riflemen Against their Right wing Under Morgan, Who Commanded in Center of the Whole, our light troops About 1000, & Can say without Ostenation that we saved the day by our timely & vigourous Attack (I believe the Indian Hoop helped A little) as we broke the Right Wing of the Enemy

---

[54] Dearborn, "A Narrative of the Saratoga Campaign – Major General Henry Dearborn, 1815," 7.
[55] Posey, "A Short Biography of the Life of Governor Thomas Posey," *Thomas Posey Papers*, Indiana Historical Society Library.

> took two 12 Pounders & one six and turned them on them.[56]

Lieutenant Colonel Wilkinson also credited Morgan's corps with routing Burgoyne's right flank:

> *True to his purpose, Morgan at this critical moment poured down like a torrent from the hill, and attacked the right of the enemy in front and flank. Dearborn at the moment, when the enemy's light infantry were attempting to change front, [to face the riflemen] pressed forward with ardour and delivered a close fire; then lept the fence, shouted, charged and gallantly forced them in disorder.* [57]

The situation was no better for Burgoyne on his left flank, where his grenadiers were decimated by General Poor's men.

Despite the collapse of his flanks, Burgoyne's center held firm. Furthermore, the commander of the British right flank, General Fraser, worked hard to restore the line. His efforts abruptly ended, however, when one of Morgan's riflemen shot him from his horse. Several years after the battle, while recounting the Battle of Saratoga to a captured British officer in Winchester, Virginia, Morgan described his role in Fraser's death. According to the unidentified British officer, Morgan declared,

---

[56] "Lt. Col. Richard Butler to Col. James Wilson, January 22, 1778," Gratz Collection, Case 4, Box 11, Historical Society of Pennsylvania.
[57] Wilkinson, 268.

> *Oh we whopped them tarnation well, surelie, said* [Morgan], *rubbing his hands; though to be sure they gave us tough work too…Me and my boys attacked a height that day, and drove Ackland and his grenadiers, but we were hardly on the top when the British rallied, and came on again with such fury that nothing could stop them. I saw that they were led by an officer on a grey horse – a devilish brave fellow; so when we took the height a second time, says I to one of my best shots, says I, you get up into that there tree, and single out him on the white horse. Dang it, 'twas no sooner said than done. On came the British again, with the grey horseman leading; but his career was short enough this time. I jist tuck my eyes off him for a moment, and when I turned them to the place where he had been – pooh, he was gone!* [58]

General Fraser, mortally wounded, was removed to the rear and died the next day in the British camp.

The loss of General Fraser and the growing pressure on their front and flanks soon proved too great for the British center and they joined the rest of Burgoyne's detachment in retreat. Eight British cannon and scores of men were abandoned on the field. Lt. Colonel Wilkinson described the carnage.

> *The ground which had been occupied by the British grenadiers presented a scene of complicated horror and exultation. In the square space of twelve or fifteen yards lay eighteen grenadiers in the agonies of death,*

---

[58] William Maxwell, ed., "A Recollection of the American Revolutionary War," *Virginia Historical Register and Literary Companion,* Vol. 6, (1853), 210.

and three officers propped up against stumps of trees, two of them mortally wounded, bleeding, and almost speechless.[59]

Most of General Burgoyne's detachment, including 300 German grenadiers who were drawn from Breymann's redoubt, retreated to the Balcarres redoubt. This bolstered the defenders there, but left Breymann's redoubt (on the extreme right of the British line) undermanned and vulnerable. Two fortified cabins between the redoubts were also weakly manned because of the failure of soldiers to return to them.

Initially these vulnerable positions were not a problem for the British because the Americans, led by General Benedict Arnold, concentrated their attack on the Balcarres redoubt. British Corporal Roger Lamb recalled,

General Arnold with a brigade of continental troops, pushed rapidly forward, for that part of the camp possessed by lord Balcarres, at the head of the British light infantry, and some of the line; here they were received by a heavy and well directed fire which moved down their ranks, and compelled them to retreat in disorder.[60]

About three hundred yards to the north, Colonel Morgan's corps prepared to storm the Breymann redoubt.

Morgan's men had advanced very close to the redoubt and used a steep hill in their front to protect them from enemy fire.[61] Lieutenant Colonel Wilkinson described the scene.

---

[59] Wilkinson, 270.
[60] Lamb, 164.
[61] Schnitzer, 50.

> *The Germans were encamped immediately behind the rail breast-work, and the ground in front of it declined in a very gentle slope for about 120 yards, when it sunk abruptly; our troops had formed a line under this declivity, and covered breast high were warmly engaged with the Germans.* [62]

Morgan was reinforced by General Learned's brigade, part of which attacked the two sparsely manned fortified cabins between the Breymann and Balcarres redoubts. Major Wilkinson was on the scene and recalled that

> *I had particularly examined the ground between the left of the Germans and the light infantry, occupied by the provincialists, from whence I had observed a slack fire; I therefore recommended to General Learned to incline to his right, and attack at that point: he did so with great gallantry; the provincialists* [defending the cabins] *abandoned their position and fled; the German flank was by this means uncovered.* [63]

Learned's brigade was joined by General Benedict Arnold, who had given up his efforts against the Balcarres redoubt and moved left towards the Breymann redoubt. Colonel Wilkinson recalled that Arnold.

> *Dashed to the left through the fire of the two lines and escaped unhurt; he then turned the right of the enemy, as I was informed by that most excellent officer,*

---

[62] Wilkinson, 272.
[63] Ibid.

> *Colonel Butler, and collecting 15 or 20 riflemen threw himself with this party into the rear of the enemy* [at Breymann's Redoubt] *just as they gave way, where his leg was broke, and his horse killed under him.* [64]

Lieutenant Colonel Richard Butler's account of the assault was similar.

> *Genl. Arnold was the first who Entered,* [Breymann's Redoubt] *one Major Morris with about 12 of the Rifle men followed him on the Rear of their Right Flank while I led up the rest of the Riflemen in front. I was the 3rd officer in* [the redoubt].[65]

Major Dearborn's light infantry also participated in the assault of Breymann's Redoubt. He described it in his memoirs:

> *The assault was commenced by the advance of Arnold with about 200 men through a cops of wood which covered the Enemies right, the appearance of Arnold on the right was the signal for us to advance and assault the front. The whole was executed in the most spirited and prompted manner and as soon as the Enemy had given us one fire, he fell back from his work to his line of tents, and as we entered he gave way and retreated in confusion.*[66]

---

[64] Ibid., 272.
[65] "Lt. Col. Richard Butler to Col. James Wilson, January 22, 1778," Gratz Collection, Case 4, Box 11, Historical Society of Pennsylvania.
[66] Dearborn, "A Narrative of the Saratoga Campaign – Major General Henry Dearborn, 1815," 8.

Whether by design or chance, the assault on Breymann's redoubt was masterfully executed and the Germans were quickly overwhelmed. General Burgoyne's line was breeched, and only nightfall saved the British from further disaster.

General Gates was overjoyed with the day's results, reportedly declaring to Colonel Morgan upon his return to camp

> *Morgan you have done wonders for your country, if you are not promoted I will not serve a day longer myself!*[67]

The American army's work was not yet finished, however. Although General Burgoyne's advance on Albany was clearly over, it remained to be seen whether his retreat to Fort Ticonderoga would succeed.

Retreat and Surrender

Under cover of darkness, General Burgoyne withdrew his army across the Great Ravine and established a new position on a steep hill overlooking the Hudson River. The position was called the Great Redoubt, and its location allowed Burgoyne to consolidate his troops and protect the river transports and hospital.

When the Americans realized that Burgoyne had withdrawn across the ravine, they took possession of his old lines and commenced a steady, but ineffectual, bombardment. General Gates sent Morgan's light corps around Burgoyne's position to

---

[67] Reverend William Hill, "Sketch of Daniel Morgan: Notes and Sermon," Virginia Historical Society, (Unpublished), 2

reconnoiter the enemy's rear and harass them. Major Dearborn participated in this reconnaissance:

> *This morning* [Oct. 8] *the Rifle men & Light Infantry & several other Regiments march'd in the Rear of the Enimy Expecting they ware Retreeting But found they ware Not. there has Been scurmishing all Day...a Large Number of the Enimy Deserted to us to Day.*[68]

General Burgoyne realized that retreat or surrender were the only options left for his army. The former was tremendously difficult, but the latter was still unthinkable. Thus, on the evening of October 8, Burgoyne began a retreat northward. Over 400 men, too injured or sick to transport, were left under a flag of truce to the care of the Americans. The rest of Burgoyne's army slowly trudged towards the village of Saratoga.

After a few miles, they halted to rest and wait for the boats to catch up. A heavy rain pelted the men all day and when they resumed their march the muddy road slowed the column to a crawl. They arrived at the heights of Saratoga after dark and collapsed on the ground in exhaustion. Lieutenant Digby described the scene.

> *We remained all night under constant, heavy rain without fires or any kind of shelter to guard us from the inclemency of the weather. It was impossible to sleep, even had we an inclination to do so from the cold and rain....*[69]

---

[68] Brown and Peckman, ed., "October 7, 1777," *Journals of Henry Dearborn*, 109.
[69] Baxter, ed., "Digby Journal," 300.

Ensign Anburey gave an equally distressing account of the British army's first night in Saratoga:

> *The army...arrived at Saratoga, in such a state of fatigue that the men had not strength or inclination to cut wood and make fires, but rather sought sleep in their wet clothes and on the wet ground.* [70]

Despite Burgoyne's slow retreat, the Americans struggled to keep pace. The rain turned the roads into a quagmire of mud, and the size of the American army, over 12,000 strong, was difficult to move in such conditions. Fortunately for the Americans, Burgoyne's retreat ceased at Saratoga.

Over the next few days, as General Burgoyne grappled with his situation, Colonel Morgan's light corps constantly harassed them. A steady American artillery bombardment added to their discomfort. By October 14th, General Burgoyne and his army had had enough. With his officer's consent, Burgoyne asked Gates for terms of surrender. General Gates was generous in his demands and on October 17th, General Burgoyne formally surrendered his army.

The most decisive battle of the Revolutionary War to date was over, and Colonel Daniel Morgan and his riflemen had played a crucial role in the victory. Yet, in General Gates's dispatch to Congress announcing Burgoyne's surrender, no mention was made of Colonel Morgan. It appears that Colonel Morgan had fallen out of favor with General Gates soon after Burgoyne's surrender. Morgan's friend, the Reverend William Hill, revealed the likely cause for this rupture as it was related to him by Daniel Morgan many years after the battle:

---

[70] Jackman, ed. "Anburey's Journal," 190.

> *Immediately after the surrender of the British army Gates took Morgan aside & apparently in confidence asked Morgan, if he knew that the greatest discontent prevailed in the American army at the Commander in chief* [General Washington] *& that many of the most valuable officers threatened to resign if a change did not take place. Morgan, expecting that Gates meant to make use of the present time, when the recent surrender of Burgoyne's army to him would give him such eclat with Congress, to move the removal of Washington in hopes of getting the place himself, & knowing how little credit was due Gates, who in both days action was not out of his strongly fortified camp replied, 'That he had one favour to ask of him which was never to mention that destestable subject to him again, for under no other man than Washington would he serve as commander in chief.' & suddenly left Gates. From that time all intimacy between them ceased.*[71]

Reverend Hill's recollection also revealed the high esteem that the British officer corps had of Morgan. General Gates hosted a dinner for the British officers soon after their surrender to which Colonel Morgan was conspicuously absent (he was apparently not invited by Gates). Military affairs, however, caused Morgan to seek General Gates at the dinner, after which he departed. Morgan recalled to Reverend Hill that,

---

[71] Hill, "Sketch of Daniel Morgan: Notes and Sermon," 2.

> *The British officers not being introduced to Col. Morgan enquired who he was, & being informed, rose from their seats at table, followed him to the door, & introduced themselves to him, so high an opinion had they conceived of him from their acquaintance they had formed with him on the field of battle.*[72]

Colonel Morgan and his riflemen had most certainly lived up to their reputation and high esteem at Saratoga. With the threat of Burgoyne eliminated, it was time for Morgan and his rifle corps to return to General Washington and the main army in Pennsylvania. They marched south within days of Burgoyne's surrender.

---

[72] Ibid.

Pennsylvania 1777

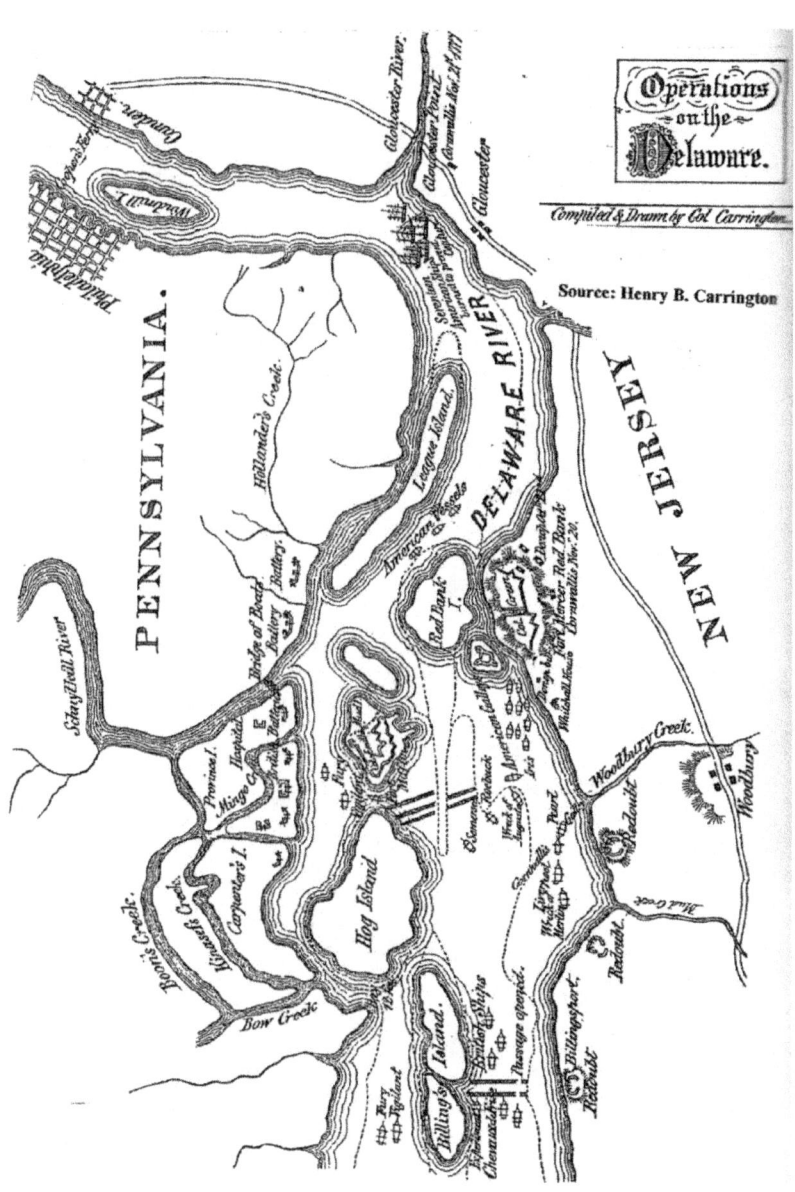

Chapter Ten

Germantown to Valley Forge

October 1777 -- February 1778

Just a few days before the victorious second battle of Saratoga in New York, General Washington also tried to score a victory against the British in Philadelphia. He targeted the 8,000 British troops posted at Germantown, just to the north of Philadelphia, and believed that if his 11,000 man army could catch them by surprise like they did ten months earlier at Trenton, the Americans could turn the direction of the war in their favor.[1]

Washington's plan was bold and complex. The army would march through the night in four columns along four different routes and converge on the enemy at sunrise in Germantown. The militia brigades, which accounted for approximately 3,000 of his men) were instructed to strike the left and right flank and rear of the enemy while the bulk of Washington's army, divided into a right and left wing, would strike the enemy head on.[2]

General Sullivan commanded the right wing of Washington's army. His force comprised his own division of Maryland troops along with General Wayne's Pennsylvania division. General Greene commanded the left wing of the army

---

[1] Thomas J. McGuire, *The Philadelphia Campaign: Germantown and the Roads to Valley Forge,* Vol. 2, (Stackpole Books, 2007), 49.
[2] Chase and Lengel, eds., "General Orders for Attacking Germantown, Oct. 3, 1777," *The Papers of George Washington, Revolutionary War Series,* Vol. 11, 375.

which included his division (of Muhlenberg's and Weedon's brigades) and General Stephen's division (of Woodford's and Scott's Virginia brigades). A recently arrived brigade of Connecticut continentals under General Alexander McDougall also marched with Greene's force, but they were to file off and attack the enemy's right flank (with the militia) as soon as the fighting began.[3]

General Washington attempted to inspire his army prior to the attack with a challenge. Noting the recent success of the American northern army in its first battle near Saratoga, Washington challenged his troops.

> *This army – the main American Army – will certainly not suffer itself to be out done by their northern Brethren – they will never endure such disgrace... Covet! My Countrymen, and fellow soldiers! Covet! A share of the glory due to heroic deeds! Let it never be said, that in a day of action, you turned your backs on the foe – let the enemy no longer triumph.... Will you suffer the wounds given to your Country to go unrevenged? Will you resign your parents – wives – children and friends to be the wretched vassals of a proud, insulting foe? And your own necks to the halter?...Every motive that can touch the human breast calls us to the most vigorous exertions – Our dearest rights – our dearest friends – our own lives – honor – glory, and even shame, urge us to the fight – And my fellow soldiers! When an opportunity presents,*

---

[3] Chase and Lengel, eds., "General Orders for Attacking Germantown, Oct. 3, 1777," *The Papers of George Washington, Revolutionary War Series*, Vol. 11, 375.

be firm, be brave, shew yourselves men, and victory is yours.[4]

General Muhlenberg's specific orders to his brigade, which were really just a reiteration of Washington's orders to the general officers, were far less emotional.

Officers of Regts. Are to see that their men have three Days Provisions Cook'd (this day included). Their men are likewise to be furnish'd with 40 Rounds of Cartridges per Man ...Flints, ect. in the best order.[5]

General Washington's plan called for a multi-pronged attack on the enemy's outposts north of Germantown at 5 a.m. General Greene's wing had one of the longest routes to march to get into position, so they commenced their march at 6 p.m. The troops left their packs and baggage in camp and were given white slips of paper to put in their hats to better distinguish themselves from the enemy in the dark (assuming it was still dark when they attacked). As there was no moon that evening, the men struggled in the dark over bad roads.[6]

By 5 a.m. three of Washington's four columns were in position to attack. General Greene's column, however, had fallen behind and was late. That didn't stop General Sullivan with the right wing of Washington's army from proceeding as

---

[4] Chase and Lengel, eds., "General Orders, October 3, 1777," *The Papers of George Washington, Revolutionary War Series*, Vol. 11, 373-374.
[5] "Muhlenberg Brigade Orders, October 3, 1777," "Orderly Book of Gen. John Peter Gabriel Muhlenberg, March 26-December 20, 1777," *The Pennsylvania Magazine of History and Biography*, Vol. 35, (1911), 63.
[6] McGuire, *The Philadelphia Campaign: Germantown and the Roads to Valley Forge,* Vol. 2, 52-53.

planned. He approached General Howe's light infantry outposts, which were situated about two miles north of the British main camp in Germantown, just after 5 a.m. A thick fog coupled with the dim light of dawn obscured visibility, but General Sullivan had vastly superior numbers and he pressed on in the face of a scattering of enemy musket fire from the British picket line and cannon fire from two 6 pound cannon.[7]

The British 2nd light battalion was now thoroughly alarmed, but they were also significantly outnumbered and had no chance of stopping Sullivan's men. The British gave ground and retreated to Benjamin Chew's large stone mansion, called Cliveden, where the 40th Regiment (300 strong) was posted. Unable to stop the American advance, more than half of the 40th Regiment retreated with the light infantry back towards the main British camp.[8]

Importantly, however, about 100 soldiers of the 40th Regiment, including their commander, Lieutenant Colonel Thomas Musgrave, barricaded themselves inside Cliveden and refused a call to surrender (actually mortally wounding the flag bearer who approached to seek their surrender).[9]

General Washington, who had accompanied General Sullivan's wing in the attack, came upon the scene and considered bypassing the obstinate enemy in the mansion, but General Knox argued that it would be dangerous to leave them in the rear and convinced Washington to order an attack.[10]

Hundreds of Washington's troops now focused their attention and efforts on dislodging 100 British troops from

---

[7] Ibid, 67.
[8] McGuire, *The Philadelphia Campaign: Germantown and the Roads to Valley Forge,* Vol. 2, 68, 70, 76-77, 80-81.
[9] Ibid, 81, 83, 85.
[10] Ibid, 87.

Cliveden, instead of joining the rest of Sullivan's wing in pushing on with the attack. Efforts to blast the British out with cannon, storm the mansion with infantry, and even burn the building to the ground, all failed. The 40th Regiment refused to budge and in doing so, they helped undermine Washington's attack.[11]

It was during these efforts to capture Cliveden that some of General Greene's troops arrived on the scene, 40 minutes behind schedule. General Woodford's brigade, on the far right of Greene's wing, was drawn to their right by the sound of the heavy fighting at Cliveden and when they reached the back of the mansion they joined the effort to take the house.[12] The rest of Stephen's division marched on past Cliveden. To their left marched the rest of General Greene's wing, all deployed into a battle line.

General Wayne, who had advanced several hundred yards past Cliveden with Sullivan's wing, heard the intense fighting to his rear and became worried that the enemy had somehow gotten behind him through the fog and smoke. He turned his brigade around and marched back towards Cliveden to investigate.[13] To his front appeared a long line of dark figures, mostly obscured by the thick fog. Suddenly a volley erupted from this line, blasting the Pennsylvanians. Those of General Wayne's troops who didn't panic naturally returned fire, striking some of General Greene's Virginians from Stephen's division.[14] A tragic case of friendly fire had occurred and the confusion of the moment caused many of the Pennsylvanians

---

[11] Ibid, 88-91.
[12] McGuire, *The Philadelphia Campaign: Germantown and the Roads to Valley Forge,* Vol. 2, 94-95.
[13] Ibid, 97.
[14] Ibid, 99.

to assume they had been flanked by the enemy. As a result, General Wayne's brigade lost its order and began to retreat.[15]

Although the mishap between the Virginians and Pennsylvanians blunted the momentum of Washington's attack, the bulk of General Greene's troops pressed forward and engaged the British 1st Light Infantry Battalion (which up to this point had not been engaged). A British officer recalled,

> *The Rebels moving on, lined a Bank & Rail under cover of the Fog, & threw in a most severe fire upon the 4th Regt...which knoc'd down almost the whole of their right wing.*[16]

Sixteen year old Joseph Plum Martin, with General McDougall's Connecticut troops who were alongside General Greene's Virginians, recalled that,

> *The enemy were driven quite through their camp. They left their kettles, in which they were cooking their breakfast, on the fires, and some of their garments were lying on the ground, which the owners had not time to put on.*[17]

Martin and his comrades in Greene's left wing briefly relished their apparent victory over General Howe's advance troops. General Muhlenberg's regiment of 9th Virginians

---

[15] Ibid.
[16] McGuire, *The Philadelphia Campaign: Germantown and the Roads to Valley Forge,* Vol. 2, 101.
[17] Joseph Plum Martin, *Private Yankee Doolittle: Being a Narrative of Some of the Adventures, Dangers and Sufferings of a Revolutionary Soldier,* (Eastern Acorn Press, 1998), 73.
    Originally published in 1840.

(commanded by Colonel George Mathews) had pushed further than any of General Greene's troops, all the way to the Market House in the center of Germantown.[18] They soon discovered, however, that their accomplishment left them vulnerable. British reinforcements swarmed upon them, determined to push the Americans back, and with no support on their flanks, the 9th Virginia was surrounded and their wounded commander forced to surrender his entire regiment.[19]

The situation was little better for the rest of General Greene's wing. With most of General Sullivan's right wing either halted or in retreat, General Greene had little chance of victory on his own. To save his troops he grudgingly ordered a withdrawal.[20]

General Washington was stunned at these developments. In a candid letter to his brother John Augustine two weeks after the battle, Washington admitted that he was still at a loss to identify the cause of their loss at Germantown.

> *After they* [the British] *had crossed* [into Philadelphia] *we took the first favourable opportunity of attacking them – This we attempted by a Nights March of fourteen Miles to surprise them (which we effectually did) so far as reaching their Guards before they had notice of our coming, and but for a thick Fog rendered so infinitely dark at times, as not to distinguish friend from Foe, at the distance of 30*

---

[18] McGuire, *The Philadelphia Campaign: Germantown and the Roads to Valley Forge,* Vol. 2, 114-115.

[19] McGuire, *The Philadelphia Campaign: Germantown and the Roads to Valley Forge,* Vol. 2, 115.

[20] Ibid, 115-116.

> *Yards, we should, I believe, have made a decisive & glorious day of it.*[21]

General Washington continued.

> *But Providence – or some unaccountable something, designd it otherwise; for after we had driven the Enemy a Mile or two, after they were in the utmost confusion, and flying before us in most places, after we were upon the point (as it appeard to every body) of grasping a compleat Victory, our own Troops took fright & fled with precipitation and disorder. How to account for this I know not, unless, as I before observ'd, the Fog represented their own Friends to them for a Reinforcement of the Enemy as we attacked in different Quarters at the same time, & were about closing the Wings of our Army when this happened.*[22]

General Washington added that the fighting lasted nearly three hours during which many of his troops expended all of their ammunition.[23] With much of his army retreating on their own in a disorganized manner, General Washington ordered them to return to their old encampment at Pennibackers Mill (about 20 miles from Philadelphia) where they could treat the wounded and reorganize themselves.

---

[21] Chase and Lengel, eds., "General Washington to John Augustine Washington, October 18, 1777," *The Papers of George Washington, Revolutionary Series*, Vol. 11, 551.

[22] Chase and Lengel, eds., "General Washington to John Augustine Washington, October 18, 1777," *The Papers of George Washington, Revolutionary Series*, Vol. 11, 551.

[23] Ibid.

Awaiting Howe's Next Move

American losses at Germantown were significant and American morale suffered as a result. The army gradually reformed at Pennypacker's Mill on Perkiomen Creek (about 20 miles from Philadelphia) where Washington informed Congress that

> *My intention is, to encamp the Army at some suitable place, to rest and refresh the Men, and recover them from the still remaining effects of that disorder naturally attendant on a Retreat. We shall here wait for the Reinforcements coming on, and shall then act according to circumstances.*[24]

Washington needed the reinforcements that were marching from General Gates's army in New York because he estimated that he lost approximately 1,000 troops at Germantown in killed, wounded, and missing. He speculated that not all of the missing were captured, some, he felt, had deserted in the fog of battle.[25]

General Muhlenberg's brigade suffered the highest losses in the army at Germantown, losing nearly the entire 9th Virginia Regiment to captivity. To offset this loss, General Washington attached the 1st Virginia State Garrison Regiment, approximately 226 effectives, to Muhlenberg's brigade.[26]

---

[24] Chase and Lengel, eds., "General Washington to John Hancock, October 7, 1777," *The Papers of George Washington, Revolutionary War Series*, Vol. 11, 417.

[25] Chase and Lengel, eds., "General Washington to John Augustine Washington, October 18, 1777," *The Papers of George Washington, Revolutionary War Series*, Vol. 11, 552.

[26] Chase and Lengel, eds., "General Orders, October 7, 1777," *The Papers of George Washington, Revolutionary War Series*, Vol. 11, 415.

This regiment was raised the previous year as one of three state garrison regiments to offset the absence of the many Virginia continental regiments that had left the state. Virginia's leaders wanted more than just militia available to defend the commonwealth, so three garrison regiments were formed in 1776. Soldiers in these regiments served full time for three years with the understanding that they would not be ordered out of the state.[27]

This was not an absolute restriction however, and in the fall of 1777 Governor Patrick Henry and the Virginia Assembly responded to appeals for more assistance from General Washington and Congress and ordered the 1st Virginia State Garrison Regiment northward to join Washington's army.[28] The regiment was commanded by Colonel George Gibson and arrived just prior to the Battle of Germantown.

In the weeks following the Battle of Germantown, Washington's army struggled to address long neglected issues such as supply shortages and military discipline. Brigade commanders were instructed to personally supervise the drafting of accurate troop returns as well as lists of missing clothing items that were needed to properly outfit each soldier in their brigades.[29] Four days after General Washington requested these returns, with nine regiments not reporting as ordered, the totals from those units that did submit returns showed that Washington's army needed, *"3,084 coats, 4,051 waistcoats, 6,148 breeches, 8,033 stockings, 6,472 shoes,*

---

[27] Hening, ed., *Statutes at Large*, Vol., 192-194.
[28] Chase and Lengel, eds., "Patrick Henry to General Washington, September 5, 1777, Footnote 3," *The Papers of George Washington, Revolutionary War Series*, Vol. 11, 152.
[29] Chase and Lengel, eds., "General Orders, October 9, 1777," *The Papers of George Washington, Revolutionary War Series*, Vol. 11, 452-453.

6,330 shirts, 137 hunting shirts, 4,552 blankets, 2,399 hats, 341 stocks, 356 overalls, and 1,140 knapsacks.[30]

Along with struggling with the chronic supply shortage in the army, General Washington used the weeks after Philadelphia's fall to address several disciplinary issues among his general officers. Accusations of misconduct in some form during the recent campaign were leveled against Generals Sullivan, Maxwell, Wayne, and Stephen.

Many of Virginia's general officers spent a sizable amount of time on boards of inquiry and courts martial to address the charges brought against their fellow officers. The only senior officer found guilty of significant wrongdoing was General Adam Stephen; he was convicted of unofficerlike behavior during the retreat from Germantown and of also being too often intoxicated while in the service, *"to the prejudice of good order and military discipline."*[31] General Washington approved the sentence of the court for General Stephen -- dismissal from the army -- and it was so ordered. Stephen's departure left General Muhlenberg as the ranking field officer from among the Virginian brigadiers, but this status would prove to be short lived.

Not all of the news in October was bad for the American army or American cause. Hopeful news arrived in mid-October of a decisive American victory in a second battle at Saratoga. This was followed just days later with even better news, General John Burgoyne's entire British army at Saratoga had

---

[30] Chase and Lengel, eds., "General Washington to John Hancock, October 13-14, 1777, Enclosed Return," *The Papers of George Washington, Revolutionary War Series*, Vol. 11, 500.

[31] Frank E. Grizzard, Jr. and David R. Hoth, eds., "General Orders, November 20, 1777," *The Papers of George Washington, Revolutionary War Series,* Vol. 12, (University Press of Virginia, 2002), 327-328.

surrendered to the American northern army under General Horatio Gates. This significant victory took some of the sting out of the recent defeats at Brandywine and Germantown (not to mention the loss of Philadelphia) and boosted American morale throughout the country.

Delaware River Forts

While General Washington welcomed the news about Saratoga, he also focused on a long shot opportunity to force the British army from Philadelphia by blocking their supply line on the Delaware River. Two forts on the banks of the river a few miles south of Philadelphia (Fort Mifflin in Pennsylvania and Fort Mercer in New Jersey) anchored American efforts to stop the British navy from reaching Philadelphia. The primary purpose of these two forts was to defend the water obstacles that blocked the channel in the Delaware River. As long as these forts were in American hands, gunfire from their cannon made it impossible for the British navy to dismantle and overcome the underwater obstacles in the river. Every day that the British navy was delayed created a greater logistical challenge for General Howe and his army (who needed the supplies aboard the blocked British transport ships in the Delaware River).

General Washington realized the significance of holding the forts and sent the 1st and 6th Virginia Regiments under Lieutenant Colonel John Green, just 200 strong, to reinforce the garrison in Fort Mifflin in mid-October.[32] They joined Major

---

[32] "Division Orders, October 17, 1777, Orderly Book of General Muhlenberg, March 26 – Dec. 20, 1777," *Pennsylvania Magazine of History and Biography*, Vol. 35, No. 1, (1911), 84. and
Chase and Lengel, eds., "Instructions to Lieutenant Colonel John Green, October 18, 1777," and "General Washington to Colonel Christopher

Robert Ballard, who commanded a small detachment of Virginians drawn from several Virginia regiments already at the fort.[33] Lieutenant Colonel Samuel Smith of Maryland commanded the garrison at Fort Mifflin.

Duty at Fort Mifflin proved very hard on the troops posted there. While the British built several artillery batteries on the Pennsylvania shoreline, several warships shelled the fort. Lieutenant Colonel Smith described one bombardment that blew up one of the barracks, but would have done even more damage if not for the actions of a Virginia officer.

Yesterday [October 19] *a red hot ball entered our Laboratory, where were two boxes of ammunition (about 30 cartouches) which blew up the barracks, and had it not been for the activity of Capt. Wells of the 4th Virginia, and Capt. Luct, in putting out the fire, would have done much damage.*[34]

A week later, with the arrival of Lieutenant Colonel Green's detachment of 1st and 6th Virginians, Lieutenant Colonel Smith updated General Washington of the condition of the troops.

Green, October 18, 1777," *The Papers of George Washington, Revolutionary War Series*, Vol. 11, 543.

[33] Chase and Lengel, eds., "Lieutenant Colonel Samuel Smith to General Washington, October 20, 1777," *The Papers of George Washington, Revolutionary War Series*, Vol. 11, 565.

[34] Chase and Lengel, eds., "Lieutenant Colonel Samuel Smith to General Washington, October 20m 1777," *The Papers of George Washington, Revolutionary War Series*, Vol. 11, 565-566.

> *Fifty Blankets as many pair of Shoes, 4 Coats, 1 Vest, 4 pair Pants & two Great Coats (all farmers) were all I received this day for my poor ragged fellows, now chiefly without Breeches, who are oblig'd to turn out before day, & perhaps may Soon be oblig'd to be So all Night, the last reinforcement* [Green's Virginians] *are equally unfurnish'd. The Garrision ought to be well Cloth'd or we destroy their Constitutions.*[35]

Lieutenant Colonel Smith reported to General Washington on November 9, that despite the continued sporadic bombardment of the British, *"we shall have put the Fort in a good posture of defence."*[36] Smith added a request that once the fort's defenses had reached that stage,

> *The Officers of the Virginia Regiments and my party hope your Excellency will relieve them and their men. Your Excellency will see the propriety of this request when I assure you that out of 200 Men completely Officer'd which my party consisted of, there are not now in Garrison more than 4 Officers and 65 privates. The 6th Virginia Regt. brought 120 rank and file, and this morning returned only 46 fit for duty, the* [1st Virginia] *nearly in proportion….*[37]

---

[35] Grizzard, Jr. and Hoth, eds., "Lieutenant Colonel Samuel Smith to General Washington, October 26, 1777," *The Papers of George Washington, Revolutionary War Series,* Vol. 12, 23.

[36] Grizzard, Jr. and Hoth, eds., "Lieutenant Colonel Samuel Smith to General Washington, November 9, 1777," *The Papers of George Washington Revolutionary War Series,* Vol. 12, 184.

[37] Ibid.

Smith noted that, *"there has not been one night without One, two or three Alarms, One half the Garrison are constantly on fatigue* [repairing damage from the steady bombardment] *and guard.*[38]

The situation grew much worse the following day when British shore batteries joined the warships in a constant bombardment. Lieutenant Colonel Smith described the damage down to the fort by the relentless bombardment.

> *Their Shot...rakes the Pallisades* [wooden wall] *fronting the Meadow, and Cuts down 4 or 5* [posts] *at a time, they have laid open a great part of that side, and chiefly destroyed that range of Barracks...*[their] *incessant fire...have dismounted 3 of our Blockhouse Guns, and much injured the Block houses and the other Range of Barracks. We cover our Men under the Wall, and have the good fortune as yet to escape unhurt; in 5 or 6 Days (unless the Siege can be rais'd) the fort will be laid open, and everything destroyed, if they continue to Cannonade and Bombard us as they have done.... Our men already half Jaded to Death with constant fatigue, will be unfit for service.*[39]

The next day Lieutenant Colonel Smith was nearly killed when a British cannonball crashed through the fireplace of his quarters, striking his hip and bringing much of the chimney

---

[38] Ibid.
[39] Grizzard, Jr. and Hoth, eds., "Lieutenant Colonel Samuel Smith to General Washington, November 10, 1777," *The Papers of George Washington, Revolutionary War Series*, Vol. 12, 203-204.

down upon him. He was evacuated from the island to Fort Mercer where it was determined his injuries were not serious.[40]

The Virginian troops in the fort, whom Smith described the day after his injury as, *"worn out with fatigue watchings & Cold,"* were also removed from the fort by November 14th, replaced by 400 troops from New England.[41] Their stay was brief, for on the evening of November 15th, with the fort virtually demolished, the garrison was finally evacuated.[42]

Lieutenant Colonel Smith, ever considerate of his men, wrote one last letter to General Washington to advocate for them and praise some of the Virginians under his command.

> *My party taken (as your Excelly knows) from the picquet, think they have done their Tour of duty, & hope for your Excellys permission to join their respective Regiments, who (they say) want their immediate Attention. The Officers have no Cloths with them.... Capt. Dickinson of the first Virga. Regt. deserves much Attention. He stay'd with & assisted Fleury, he is a brave industrious good Officer. Capt. Walls of the 4th Virga has distinguish'd himself on every Occasion, for a brave, Industrious and prudent Officer.*[43]

---

[40] Grizzard, Jr. and Hoth, eds., "Lieutenant Colonel Smith to General Washington, November 12, 1777," *The Papers of George Washington, Revolutionary War Series,* Vol. 12, 231.

[41] Ibid, and Grizzard, Jr and Hoth., eds., "General Varnum to General Washington, November 14, 1777," *The Papers of George Washington, Revolutionary War Series,* Vol. 12, 261.

[42] Grizzard, Jr. and Hoth, eds., "General Varnum to General Washington, November 16, 1777," *The Papers of George Washington, Revolutionary War Series,* Vol. 12, 283.

[43] Grizzard, Jr. and Hoth, eds., "Lieutenant Colonel Smith to General Washington, November 16, 1777," *The Papers of George Washington, Revolutionary War Series,* Vol. 12, 281-282.

Operations in New Jersey

General Washington had closely monitored the situation with the river forts and pondered the idea of attacking Philadelphia to help relieve pressure on them but accepted the advice of a war council that voted against an attack and instead, posted the army at Whitemarsh (about 20 miles north of Philadelphia).[44] It was hoped that General Howe would interpret this movement as another threat to Philadelphia, which might then dissuade him from sending additional troops against the river forts.

The war council also recommended that sufficient reinforcements be sent to the river forts to keep them at full strength. A period of relative inactivity settled over the main American army as General Washington waited for General Howe to resume operations against him or go into winter quarters.

Although the loss of Fort Mifflin was discouraging, General Washington maintained hope that Fort Mercer, which was across the river from Fort Mifflin and still in American hands, might, by itself, be able to stop the British navy from passing the river obstacles. Unfortunately for the Americans, the commander of Fort Mercer, Colonel Christopher Greene of Rhode Island, informed Washington that British ships had found a passage up the river past Fort Mifflin and he feared his garrison was in danger of being besieged and captured.[45]

---

[44] Grizzard, Jr. and Hoth, eds., "Council of War, October 29, 1777," *The Papers of George Washington, Revolutionary War Series,* Vol. 12, 46-48.

[45] Grizzard, Jr. and Hoth, eds., "Colonel Christopher Greene to General Washington, November 17, 1777," *The Papers of George Washington, Revolutionary War Series,* Vol. 12, 288.

General Washington was not ready to give up Fort Mercer, however, and ordered several general officers in camp to proceed to the fort to evaluate whether an effort to keep it should be attempted.[46] They reported back on November 19th, that they thought it was both feasible and important to keep possession of Fort Mercer.[47]

General Washington acted immediately and ordered General Greene to cross the Delaware River with his division (Muhlenberg's and Weedon's brigades) and take command of all of the reinforcements sent to New Jersey to assist Fort Mercer.[48]

> *Very much will depend upon keeping possession of Fort Mercer,* [explained Washington] *as to reduce it, the Enemy will be obliged to put themselves in a very disagreeable position to them and advantageous to us, upon their Rear. Therefore desire Colo. Green to hold* [Fort Mercer] *if possible till the relief arrives.*[49]

Colonel Daniel Morgan's rifle corps, which had arrived at Washington's camp at Whitemarsh in mid-November significantly understrength, was also attached to General Greene's detachment. The toil of the Saratoga campaign and

---

[46] Grizzard, Jr. and Hoth, eds., "Orders to Major Generals Arthur St. Clair and Johann Kalb and Brigadier General Henry Knox, November 17, 1777," *The Papers of George Washington, Revolutionary War Series,* Vol. 12, 298.

[47] Grizzard, Jr. and Hoth, eds., "General Washington to General Varnum, November 19, 1777," *The Papers of George Washington, Revolutionary War Series,* Vol. 12, 322-323.

[48] Grizzard, Jr. and Hoth, eds., "General Washington to General Varnum, November 19, 1777," *The Papers of George Washington, Revolutionary War Series,* Vol. 12, 323.

[49] Ibid.

the long march from New York took a heavy toll on Morgan's men. *"There are not more than one hundred and Seventy of Morgan's Corps fit to march, as they in general want Shoes,"* noted General Washington upon their return.[50]

Unfortunately, it was too late to save Fort Mercer. On the night of November 19th, Colonel Christopher Greene, who had valiantly defended Fort Mercer against a Hessian assault in October, feared that his garrison was about to be invested and trapped by the enemy and opted instead to save his troops by evacuating the fort.[51] General Greene, who had crossed the Delaware River above Philadelphia at Burlington, New Jersey with his division, informed Washington of the news.[52]

There was still a chance for General Greene to inflict some loss on the large enemy force that had crossed into New Jersey to attack Fort Mercer. Greene had more than just his division of Virginians with him. General Varnum's Rhode Island brigade and General Jedediah Huntington's Connecticut brigade were in New Jersey as were the remnants of Colonel Morgan's Rifle Corps and Captain Henry Lee's troop of light horse. General Greene also had over a thousand New Jersey militia in the field ready to follow his command and General John Glover's brigade of Massachusetts continentals were expected to join him as well.[53]

---

[50] Grizzard, Jr. and Hoth, eds., "General Washington to General Greene, November 22, 1777," *The Papers of George Washington, Revolutionary War Series,* Vol. 12, 349-350.

[51] Grizzard, Jr. and Hoth, eds., "General Varnum to General Washington, November 20, 1777," *The Papers of George Washington, Revolutionary War Series,* Vol. 12, 336.

[52] Grizzard, Jr. and Hoth, eds., "General Greene to General Washington, November 21, 1777," *The Papers of George Washington, Revolutionary War Series,* Vol. 12, 340.

[53] Grizzard, Jr. and Hoth, eds., "General Greene to General Washington,

With this sizable force General Greene had hoped to attack General Lord Cornwallis in New Jersey, but Cornwallis had also been reinforced with British troops from New York and actually outnumbered Greene's troops with approximately 5,000 of his own.[54]

Nevertheless, General Greene defiantly remained in New Jersey after Fort Mercer was abandoned and skirmished with British foraging parties. The Marquis de La Fayette, a French volunteer in the American army, commanded a portion of Greene's troops in one such skirmish in late November. Morgan's riflemen, commanded by Lieutenant Colonel Richard Butler, comprised half of La Fayette's force and greatly impressed him. General Greene noted after the engagement that, *"The Marquis is charmed with the spirited behaviour of the Militia & Rifle Corps."* [55] La Fayette lavished praise on the riflemen in his report to General Washington:

I take the greatest pleasure to let you know that the conduct of our soldiers is above all praises – I never saw men so merry, so spirited, so desirous to go on to the enemy what ever forces they could have as the little party was in this little fight. I found the riflemen above even their reputation...I must tell too the

---

November 21, 1777," and "General Varnum to General Washington, November 21, 1777," *The Papers of George Washington, Revolutionary War Series,* Vol. 12, 340 and 343.

[54] Grizzard, Jr. and Hoth, eds., "General Greene to General Washington, November 24, 1777," *The Papers of George Washington, Revolutionary War Series,* Vol. 12, 376-378.

[55] Grizzard Jr. and Hoth, eds., "General Greene to General Washington, November 26, 1777," *The Papers of George Washington, Revolutionary War Series,* Vol. 12, 409.

> *riflemen had been the whole day running before my horse without eating or taking any rest.*[56]

General Greene marched the bulk of his force back to Whitemarsh on November 28, but Morgan's corps and a detachment of Virginian cavalry under Captain Henry Lee remained in New Jersey a while longer to bolster the local militia and harass the enemy.[57] After another week of skirmishing with the British, Morgan and his riflemen rejoined Washington's army at Whitemarsh just in time to help fend off General Howe's last major operation of the year.

Battle of Whitemarsh

On the night of December 4, over 10,000 British troops marched from Philadelphia towards the American camp at Whitemarsh.[58] The American army, about 12,000 strong, was alerted, and manned their fortifications among the hills of Whitemarsh in anticipation of an attack.[59] General Howe halted at Chestnut Hill, about three miles south of Whitemarsh to access Washington's defenses and determine his next move. During this pause, General Washington sent two detachments of Pennsylvania militia, (some 1,600 strong) and a regiment of Connecticut continentals forward towards Chestnut Hill to

---

[56] Grizzard Jr. and Hoth, eds., "General LaFayette to General Washington, November 26, 1777," *The Papers of George Washington, Revolutionary War Series,* Vol. 12, 418-419.
[57] Grizzard Jr. and Hoth, eds., "General Greene to General Washington, November 28, 1777," *The Papers of George Washington, Revolutionary War Series,* Vol. 12, 428.
[58] David Martin, *The Philadelphia Campaign, June 1777 – July 1778,* (Da Capa Press, 1993), 160.
[59] Lesser, "A General Return of the Continental Army…Dec. 3, 1777," 53.

skirmish with Howe's advance parties and obstruct his reconnaissance.[60] Unfortunately, despite a determined but brief stand by the Connecticut continentals, Washington's skirmishers were easily dispersed by General Howe's troops. Yet, the British commander remained stationary for the next two days, reluctant to attack the strong American position in the hills to his front, but determined to strike one last blow against the rebels.

General Howe made his move early in the morning of December 7. Under cover of darkness he shifted his army several miles to the east in two columns in an effort to flank Washington.[61] General Cornwallis commanded the main column that was to flank the Americans while General Charles Grey commanded a second column that was to distract Washington and his troops by threatening a direct frontal attack. Howe's plan was similar in design to the movements that brought him victory over Washington at Long Island in 1776 and Brandywine in September. This time, however, the American commander anticipated Howe's movement and placed Colonel Morgan's rifle corps, supported by a detachment of Maryland militia, on his far left flank to guard against just such a move.

The engagement that ensued was fierce and costly to both sides. Captain Johann Ewald, of the German Jagers, recalled,

---

[60] Thomas McGuire, *The Philadelphia Campaign: Germantown and the Roads to Valley Forge*, Vol. 2. (Stackpole Books, 2007), 241.

[61] Grizzard Jr. and Hoth, eds., "General Washington to Patrick Henry, December 10, 1777," *The Papers of George Washington, Revolutionary War Series*, Vol. 12, 590.

> *The light infantry fell into an ambuscade which the American Colonel Morgan and his corps of riflemen had laid in a marshy wood, through which over fifty men and three officers were killed.*[62]

General Howe acknowledged the success of Morgan's ambush, reporting to Lord George Germain after the engagement that

> *The thickness of the wood where the rebels were posted, concealing them from the view of the light infantry, occasioned a loss of one officer killed, three wounded, and between twenty and thirty men killed and wounded from the first fire.*[63]

General Washington's brief description of the engagement to Governor Patrick Henry also noted the heavy casualties Morgan's corps inflicted on the enemy:

> *As soon as they* [Howe's troops] *began to move* [against the American left flank] *Colo. Morgan with the light Corps under his command and the Maryland Militia attacked their right flank, and I am informed did them a good deal of damage....*[64]

---

[62] Ewald, 109.
[63] Grizzard Jr. and Hoth, eds., "General Orders, December 8, 1777," Note 1 "General Howe to Lord George Germain, 13 December, 1777," *The Papers of George Washington, Revolutionary War Series,* Vol. 12, 573.
[64] Grizzard Jr. and Hoth, eds., "General Washington to Patrick Henry, December 10, 1777," *The Papers of George Washington, Revolutionary War Series,* Vol. 12, 590.

One of General Washington's aides, Major John Laurens of South Carolina, described a similar scene:

> *Upon hearing they* [the British] *were advancing in two columns Morgan's corps and the Maryland militia were ordered to harass their right flank; there was some very smart firing in consequence, between Morgan's and the British light infantry.*[65]

Major Laurens learned from Colonel Morgan himself about the effectiveness of the riflemen, adding that,

> *Col. Morgan, who has no need of boasting to establish the reputation of his corps, says the British light infantry lost a great many in their skirmish with him.*[66]

A British officer's account of the engagement confirmed the intense fire of Morgan's men. *"Their fire was more destructive for the time & number than had happened* [in] *the* [entire] *War,"* wrote Lieutenant Frederick Wetherall.[67]

These accounts of Morgan's bold conduct at Whitemarsh were contradicted by another British officer who claimed that the British light infantry commander, Colonel Robert Abercromby, routed Morgan's riflemen with an aggressive new tactic. Colonel George Hager, who served with the German Jagers, claimed:

---

[65] Dixon and Hunter, "Extract of a letter from an officer in camp, Dec. 16, 1777," *Virginia Gazette*, "December 26, 1777," 1.
[66] Ibid.
[67] McGuire, *The Philadelphia Campaign: Germantown and the Roads to Valley Forge*, Vol. 2, 251.

> *The moment* [the riflemen] *appeared before* [Abercromby], *he ordered his troops to charge with the bayonet, not one* [rifleman] *out of four, had time to fire, and those who did, had no time to load again, they did not stand three minutes;*[68]

This bold tactic of immediately charging at Morgan's riflemen upon contact with them might help account for the high casualties suffered by Morgan's corps. Twenty-seven riflemen reportedly fell at Whitemarsh, including Major Joseph Morris of New Jersey.[69] This distinguished rifle officer was struck in the head and mortally wounded and his loss was severely felt by Colonel Morgan and the rifle corps.

Although by most accounts Morgan's riflemen and the Maryland militia that were with them fought bravely, they could not hold against General Howe's column and were forced to withdraw to the main American line at Whitemarsh. Howe did not pursue, halting on Edge Hill for the night.

General Washington expected General Howe to renew his attack the next morning and used the conduct of Colonel Morgan and his rifle corps as an example for the army to emulate, declaring in the next day's general orders that

> *The Commander in Chief returns his warmest thanks to Col. Morgan, and the officers and men of his intrepid corps, for their gallant behavior in the several skirmishes with the enemy yesterday – He hopes the most spirited conduct will distinguish the*

---

[68] Ibid.
[69] Grizzard Jr. and Hoth, eds., "General Washington to Henry Laurens, December 10, 1777," *The Papers of George Washington, Revolutionary War Series,* Vol. 12, 592.

whole army, and gain them a just title to the praises of their country, and the glory due to brave men....[70]

The British failed to approach, however, and instead, marched back to Philadelphia. General Howe refused to risk his army in a direct assault on Washington's strong position. Howe's withdrawal marked the end of the 1777 campaign. The British looked forward to a relatively comfortable winter in Philadelphia. The Americans were not as fortunate.

Unlike the previous winter, in which General Washington's army was spread out among New Jersey's Watchtung Mountains, the American commander wanted to keep the army intact. To do this he needed a location that could accommodate the army's needs. Whitemarsh would not work because it was too close to Philadelphia; just a few hours march from the British army. Washington needed someplace further away (to allow for a proper warning if the British did try to attack over the winter) but also defensible and logistically sustainable. He settled on Valley Forge and marched the army there on December 19th.

Valley Forge

Valley Forge was approximately 25 miles northwest of Philadelphia and provided adequate terrain on which to resist a possible British attack. The Schuylkill River protected the American left flank and a steep hill, called Mount Joy, covered their rear. Although the front and right flank of the

---

[70] Grizzard Jr. and Hoth, eds., "General Orders, December 8, 1777," *The Papers of George Washington, Revolutionary War Series,* Vol. 12, 571.

encampment possessed few natural barriers, the open terrain made an attack from those directions very hazardous.

General Washington ordered that log huts and earthworks be constructed immediately, but it took nearly a month before the entire army was adequately sheltered. Two lines of earthworks were built. The outer line extended along a ridge from the Schuylkill River to the foot of Mount Joy.

Most of the army, including the Virginia brigades of General Muhlenberg and General Weedon, were stationed along this line in rows of huts behind the fortifications. Muhlenberg's troops were posted on the left flank of the outer line with Weedon's troops to their right. An inner defense line was built along Mount Joy. General Scott's and General Woodford's brigades were posted on the inner line.

For much of their time at Valley Forge, Washington's troops lacked both clothing and provisions, especially at the beginning of the encampment. On December 22, General James Varnum of Rhode Island reported to General Washington.

> *Three Days successively, we have been destitute of Bread. Two Days we have been intirely without Meat. –It is not to be had from Commissaries. –Whenever we procure Beef, it is of such a vile Quality, as to render it a poor Succedanium for Food. The Men must be supplied, or they cannot be commanded.* [71]

General Washington forwarded the bad news to Congress:

---

[71] Joseph Lee Boyle, "General Varnum to General Washington, December 22, 1777," *Writings from the Valley Forge Encampment of the Continental Army*, Vol. 1, (Bowie: Heritage Books Inc., 2000), 2.

> *I do not know from what cause this alarming deficiency, or rather total failure of Supplies arises: But unless more vigorous exertions and better regulations take place in that line and immediately, This Army must dissolve.*[72]

While the main army struggled with supply problems, General Washington ordered light parties to patrol south of Valley Forge and engage enemy patrols and foraging parties whenever possible. Colonel Morgan's riflemen and Captain Henry Lee and his troop of dragoons made up some of the light parties assigned to this duty.

Working in the vicinity of Radnor Meeting House, roughly seven miles southeast of Valley Forge, Lee's patrols extended south and frequently engaged enemy troops within a few miles of Philadelphia. In one incident, one of Captain Lee's dragoons was captured, but was then rescued by a party of Colonel Morgan's riflemen. Captain Lee described the incident to General Washington.

> *On hearing of the enemys excursion I immediately left camp, & moved down towards Darby. Early this morning we set out on the partisan business; having fully reconnoitered the enemys disposition, whom we found posted in force...I divided my Troop. Lt. Lindsay with Major Clark whom we accidentally met with took the route towards Chester, while myself with the other party visited the right of their encampment.*[73]

---

[72] Grizzard Jr. and Hoth, eds., "General Washington to Henry Laurens, December 22, 1777," *The Papers of George Washington, Revolutionary War Series,* Vol. 12, 667.

[73] Gizzard Jr. and Hoth, eds., "Captain Henry Lee to General Washington,

Lee reported that, *"nothing of consequence,"* happened to his party, but one of the dragoons with Lieutenant Lindsay was captured by a party of enemy horsemen. Lindsay sought the assistance of Lieutenant Colonel Richard Butler, who was nearby with a large detachment of Morgan's riflemen, and they struck back, capturing ten of the enemy horsemen and freeing Lee's captive dragoon.[74] Lee's troop further benefitted from the affair when they were allowed to keep some of the gear of the captured enemy dragoons.[75]

Captain Lee and his troop continued their patrols outside of Valley Forge into the new year and unlike the bulk of the American cavalry, which spent the winter safely in Trenton, Lee's troop spent most of January posted at a stone farmhouse (Scott's Farm) near Radnor Meeting House.[76] On January 4th, Captain Lee shared his thoughts with General Washington on what was needed to properly reconnoiter and secure the area around Radnor.

> *Agreeable to your Excellency's direction I have informed myself minutely with the country in the vicinity of Radnor meeting-house. To effect the object of your Excellency's wishes, vizt. security to the camp: I conceive it absolutely necessary to establish two*

December 23, 1777," *The Papers of George Washington, Revolutionary War Series,* Vol. 12, 689.

[74] Ibid.

[75] Grizzard Jr. and Hoth, eds., "General Stirling to General Washington, December 24, 1777," *The Papers of George Washington,* Vol. 12, 696-697.

[76] Edward G. Lengel, ed., "General Washington to General Casimir Pulaski, December 31, 1777," *The Papers of George Washington, Revolutionary War Series,* Vol. 13, (Charlottesville University Press of Virginia, 89.

> *posts of horse. The one to...patrole one mile, more or less, in advance of the advanced centinal. The other to be fixed near Newtown-square...to patrole the square-tavern. This last place, is without any guard.... It appears to me, that the post established, or to be established at the meeting-house is very far from being secure, unless great attention is paid to the square, at which place the three roads Hartford, Darby & Chester all unite.*[77]

Captain Lee explained that two troops of cavalry were needed to secure the area at night, but in the daytime, one troop of horse could handle the job. Lee went on to state the necessity of impressing provisions from local farmers and he offered to engage in the business of intelligence gathering by serving as a contact for American spies in and around Philadelphia.[78]

With nearly all of the army's cavalry posted in Trenton, there were few dragoons left at Valley Forge to send to Lee so his request for additional dragoons was ignored. Lee had to rely on Colonel Morgan's riflemen at Radnor Meeting House for support and Morgan relied on Lee's dragoons to serve as videttes on Morgan's thin piquet line.

Undoubtedly frustrated by the difficult situation he found himself, Captain Lee was pleasantly surprised to learn in mid-January that he and his officers had gained some notoriety in the *New Jersey Gazette*. The newspaper credited Captain Lee and his troop of dragoons with capturing over 125 enemy

---

[77] Lengel, ed., "Captain Henry Lee to General Washington, January 4, 1777," *The Papers of George Washington, Revolutionary War Series*, Vol. 13, 141-142.
[78] Ibid.

prisoners since the summer and called for the service of Lee and his officers to be duly rewarded.

> *A troop of dragoons in Bland's regiment, seldom having more than 25 men and horses fit for duty, has since the first of August last, taken 125 British and Hessian privates, besides four commissioned officers, with the loss of only one horse. This Gallant Corps is under the command of Captain Lee, Lt. Lindsay and Cornet Peyton whose merit and services it is hoped will not be passed unnoticed or unrewarded.*[79]

Raid on Scott's Farm

Lee's notoriety extended to the enemy, who sent a large cavalry detachment to Scott's Farm to seize Lee late in the evening of January 18th. An aide to British General William Howe noted the effort in his diary.

> *At 11 o'clock at night, 40 dragoons were detached by a long roundabout way to seize a rebel dragoon captain by the name of Lee, who has alarmed us quite often by his boldness....*[80]

Captain Johann Ewald recorded a similar observation in his journal but doubled the size of the British force.

> *Today the English Major Crewe was sent out with eighty horsemen to surprise the partisan Captain Lee,*

---

[79] *New Jersey Gazette*, "January 14, 1778".
[80] Lengel, ed., "Captain Muenchhausen's Diary," *The Papers of George Washington, Revolutionary War Series,* Vol. 13, 292-293.

> who stood with forty horse on this side of Valley Forge and constantly alarmed our outposts.[81]

The British rationale for the attack, to seize a bold American officer who had become a nuisance, was also mentioned in at least one American account of the affair. The *New Jersey Gazette* attributed the attack to General Howe's,

> *Longing to rob the Americans of this gallant young officer, whose attention in observing his motions, and address in surprising his parties perplexed him so much the last campaign.*[82]

Captain Lee was posted at Scott's Farm, about 16 miles west of Philadelphia, with Lieutenant Lindsay and a handful of dragoons. Major John Jameson of the 1st Continental Dragoons (Bland's regiment) was also at the farm paying Lee a visit. The bulk of Lee's troop was out on patrol in small parties and the handful of officers and men left at the farm were sheltered in a strong stone farmhouse.

Captain Lee described the opening of the engagement in a letter to General Washington after the affair.

> *About day break* [the enemy] *appeared, we were immediately alarm'd, & manned the doors & windows.*[83]

---

[81] Ewald, 121.
[82] Moore, "New Jersey Gazette, January 28, 1778," *Diary of the American Revolution*, Vol. 2, (New York: Charles Scribner, 1860), 10.
[83] Lengel, ed., "Captain Henry Lee Jr. to General Washington, January 20, 1778," *The Papers of George Washington, Revolutionary War Series,* Vol. 13, 292.

According to an account of the engagement in the *New Jersey Gazette*, Lee's men,

> *Scarcely had time to bolt the doors before* [the enemy] *began a smart firing into the windows, and demanded the immediate surrender of the house....* [Captain Lee refused to surrender and his men] *returned the fire from the windows with spirit, and, by showing themselves at different places, made as great an appearance of numbers as possible."*[84]

The British dragoons repeatedly tried to storm the house and were driven back each time. Frustrated, some of them began plundering the outbuildings which prompted Captain Lee to brazenly taunt the enemy commander.

> *Comrade, shame on you, that you don't have your men under better discipline. Come a little closer, we will soon manage it together!*[85]

After about a half hour the British gave up and withdrew, having suffered eight casualties.[86] Although four of Lee's dragoons on patrol were captured by the British on their way back to Philadelphia and another was captured at the farm when he tried to flee, Lee and his men emerged victorious from the engagement and were showered with praise from the public.

---

[84] Moore, "New Jersey Gazette, January 28, 1778," *Diary of the American Revolution*, Vol. 2, 10.
[85] Ewald, 121.
[86] Lengel, ed., "Captain Henry Lee Jr. to General Washington, January 20, 1778," *The Papers of George Washington, Revolutionary War Series*, Vol. 13, 293.

General Washington publicly acknowledged Lee's brave stand, thanking Lee and his men in the general orders.

> *The Commander in Chief returns his warmest thanks to Captn Lee & the Officers & men of his Troop for the Victory which by their superior Bravery and Address they gain'd over a party of the Enemys dragoons, who trusting in their numbers – and concealing their march by a circuitous road attempted to surprise them in their quarters. He has the satisfaction of informing the Army that Captn Lee's Vigilance baffled the Enemy's designs by judiciously posting his men in his quarters, although he had not a sufficient number to allow one for each window, he obliged the [enemy] disgracefully to retire after repeated but fruitless attempts to force their way into the house.*[87]

General Washington followed this glowing public praise for Lee with a private letter to the young captain that hinted at a reward to come.

> *Altho I have given you my thanks in the general Orders of this day for the late instance of your gallant behavior I cannot resist the Inclination I feel to repeat them again in this manner. I needed no fresh proof of your merit, to bear you in remembrance – I waited only for the proper time and season to shew it – these I hope are not far off.... Offer my sincere thanks to the whole of your gallant party and assure them that no one felt pleasure*

---

[87] Lengel, ed., "General Orders, January 20, 1778," *The Papers of George Washington, Revolutionary War Series,* Vol. 13, 286-287.

more sensibly, or rejoiced more sincerely for yours & their escape than Yr, Affectionate. G. Washington.[88]

As evidenced by the American reaction to the skirmish at Scott's Farm, Lee's bold stand was welcome news to an encampment struggling to feed itself at Valley Forge. Although most of the men were in huts by mid-January, adequate provisions and clothing for the troops remained a serious problem for the army.

The Virginia Line at Valley Forge

Washington had struggled with a shortage of clothing and supplies from the start of the encampment at Valley Forge, both of which accounted for a startling drop in the number of men fit for duty. General Muhlenberg's brigade of the 1st, 5th, 9th, and 13th Virginia Regiments reported 261 officers and men fit for duty on the last day of December 1777.[89] Another 508 officers and men were unable to serve at Valley Forge due to illness, furlough, or because they were on detached service.[90] General Weedon's brigade of the 2nd, 6th, 10th, and 14th Virginia Regiments was nearly double the size of Muhlenberg's brigade at 497 officers and men fit for duty. Another 876 soldiers in Weedon's brigade were unable to serve due to illness, furlough, or detached service.[91]

---

[88] Lengel, ed., "General Washington to Captain Lee, January 20, 1778," *The Papers of George Washington, Revolutionary War Series,* Vol. 13, 294.

[89] Lesser, ed., "A General Return of the Continental Army…December 31, 1777," *The Sinews of Independence: Monthly Strength Reports of the Continental Army,* 54.

[90] Ibid.

[91] Ibid.

General Woodford's brigade, which was posted on the inner line at Valley Forge, numbered just 266 officers and men fit for duty at the end of 1777. Over 1100 men were unable to serve for the same reasons as Muhlenberg's and Weedon's troops. Over 300 of these men reported a different reason for their inability to serve, lack of shoes and/or clothing.[92]

General Scott's brigade reported 445 officers and men fit for duty at the end of 1777 with nearly 700 others unable to serve for the usual reasons (including 71 who lacked sufficient clothing or shoes).[93]

Altogether the Virginians fit for duty at Valley Forge among the four Virginia brigades amounted to 1469 troops. Another 3,186 Virginians, nearly two thirds of the Virginians on the continental army's rolls, were unable to serve due to illness, furlough, lack of clothing, or detached service.[94]

The situation only worsened over the winter as provision and clothing remained scarce. Major Alexander Scammell of Pennsylvania painted a bleak picture at the height of the crisis in February.

> *A moments Opportunity presents of telling you our Distress in Camp has been infinite.... In all the Scenes since I have been in the army, want of provisions these ten Days past, has been the most distressing, [a] great part of our Troops 7 Days with only half a pound of Pork during the whole time – Our poor brave Soldiers living upon bread & water & naked exhibited a Sight exceedingly affecting to the Officers.*[95]

---

[92] Ibid.
[93] Ibid.
[94] Ibid.
[95] Joseph Lee Boyle, ed., "Alexander Scammell to Timothy Pickering,

In one of his many appeals for assistance, General Washington shared a similar observation about the troops with George Clinton of New York.

> *For some days past, there has been little less, than a famine in camp. A part of the army has been a week, without any kind of flesh, and the rest three or four days. Naked and starving as they are, we cannot enough admire the incomparable patience and fidelity of the soldiery, that they have not been ere this excited by their sufferings, to a general mutiny and dispersion. Strong symptoms, however, of discontent have appeared in particular instances; and nothing but the most active efforts everywhere can long avert so shocking a catastrophe.*[96]

The impact of such hardship weakened the entire army. By the end of March 1778, General Muhlenberg's brigade, which at full strength should have approached 3,000 troops, reported just 112 officers and men fit for duty. Weedon's brigade had 241 such officers and men, Woodford's brigade just 151 and Scott's brigade 222 officers and men fit for duty.[97]

The difficult conditions at Valley Forge, which resulted in rampant illness among the army, accounted for some of the loss, but the end of the two-year enlistments for the Virginian troops in the 1st through 9th Virginia Regiments was likely the

---

February 19, 1778," *Writings from the Valley Forge Encampment of the Continental Army*, Vol. 2 (Bowie MD: Heritage Books Inc., 2001), 50.

[96] Lengel, ed., "General Washington to George Clinton, February 16, 1778," *The Papers of George Washington, Revolutionary War Series*, Vol. 13, 552-553.

[97] Lesser, ed., "A General Return of the Continental Army…March 28, 1778," *The Sinews of Independence: Monthly Strength Reports of the Continental Army*, 60.

biggest reason for the enormous decline in the Virginia Line's numbers at Valley Forge. Several hundred troops were sent home on furlough after they agreed to re-enlist, and scores of officers were also sent home to recruit new troops to replace the hundreds of Virginia continentals who chose not to serve another term in the army.[98]

Much depended on the success of these recruitment parties and the perseverance of the troops, including hundreds of hardy Virginians, who remained in the field at Valley Forge over the winter of 1778.

---

[98] Ibid.

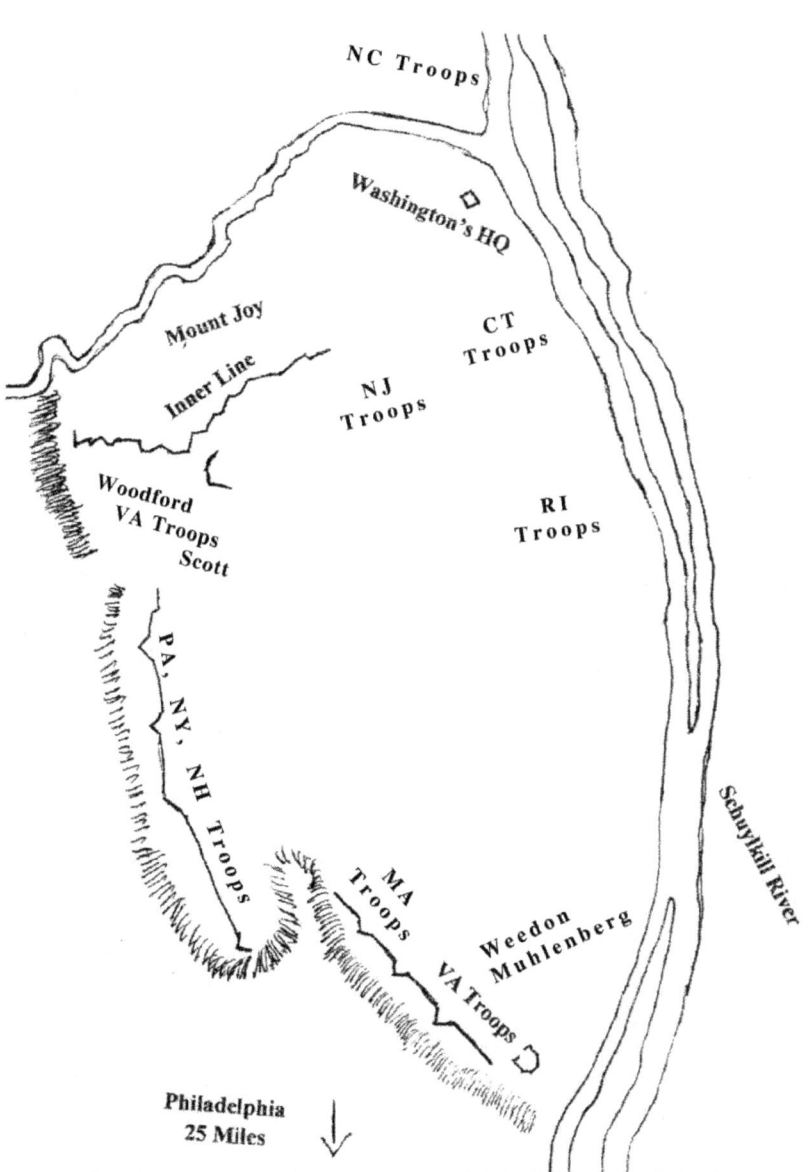

Appendix A

Virginia Continental Units 1775-78

Morgan's Independent Rifle Company   1775-1776

Stephenson's Independent Rife Company   1775-1776

1st Virginia Regiment	1775-1776*
	1776-1778
2nd Virginia Regiment	1775-1776*
	1776-1778
3rd Virginia Regiment	1776-1778
4th Virginia Regiment	1776-1778
5th Virginia Regiment	1776-1778
6th Virginia Regiment	1776-1778
7th Virginia Regiment	1776-1778
8th Virginia Regiment	1776-1778
9th Virginia Regiment	1776-1778
10th Virginia Regiment	1777-1780
11th Virginia Regiment	1777-1780

12th Virginia Regiment	1777-1780
13th Virginia Regiment	1777-1780
14th Virginia Regiment	1777-1780
15th Virginia Regiment	1777-1780
Grayson's Additional Battalion	1777-1780
1st Continental Light Dragoons	1776 ---
3rd Continental Light Dragoons	1776 ---
1st Continental Artillery Regiment	1776 ---

* Placed on Continental Establishment by the Continental Congress on December 28, 1775.

Appendix B

Engagements of Virginia Continentals in 1775-78

Siege of Boston **Summer 1775 – March 1776**
   Morgan's Rifle Company
   Stephenson's Rifle Company

Hampton **October 1775**
   One company of 2nd Virginia

Great Bridge **December 1775**
   2nd Virginia

Quebec **December 1775**
   Morgan's Rifle Company

Charlestown **June 1776**
   8th Virginia

Gwynn's Island **July 1776**
   3rd Virginia – three companies
   7th Virginia
   Virginia Continental Artillery – two companies

Harlem Heights **September 1776**
   3rd Virginia

White Plains **October 1776**
   1st Virginia
   3rd Virginia

Fort Washington November 1776
    Rawlings Rifle Corps

Retreat to the Delaware River Nov. – Dec. 1776
    1st Virginia
    3rd Virginia
    4th Virginia
    5th Virginia
    6th Virginia
    Rawlings Rifle Corps (remnants)

Trenton December 1776
    1st Virginia
    3rd Virginia
    4th Virginia
    5th Virginia
    6th Virginia
    Rawlings Rifle Corps (remnants)

2nd Trenton January 1777
    1st Virginia
    4th Virginia
    5th Virginia
    6th Virginia
    Rawlings Rifle Corps (remnants)

Princeton January 1777
    1st Virginia
    4th Virginia
    5th Virginia
    6th Virginia
    Rawlings Rifle Corps (remnants)

Boundbrook Encampment Winter-Spring 1777
    1st thru 15th Virginia Regiment
    Grayson's Additional Battalion
    1st Continental Light Dragoons

Tour of Jerseys Summer 1777
    1st thru 15th Virginia Regiment
    Grayson's Additional Battalion
    1st Continental Light Dragoons
    Morgan's Rifle Corps

Cooches Bridge September 1777
    Maxwell's Light Corps

Brandywine September 1777
    1st thru 15th Virginia Regiment
    Grayson's Additional Battalion
    1st Continental Light Dragoons
    Maxwell's Light Corps

Saratoga (Freeman's Farm) September 1777
    Morgan's Rifle Corps

Germantown October 1777
    1st thru 15th Virginia Regiment
    Grayson's Additional Battalion
    1st Continental Light Dragoons

Saratoga (Bemis Heights) October 1777
    Morgan's Rifle Corps

Fort Mifflin   November 1777
    1st Virginia
    6th Virginia
    4th Virginia – detachment

White Marsh   December 1777
    1st thru 15th Virginia Regiment
    Grayson's Additional Battalion
    1st Continental Light Dragoons
    Morgan's Rifle Corps

Valley Forge   Dec. 1777 – June 1778
    1st thru 15th Virginia Regiment
    Grayson's Additional Battalion
    1st Continental Light Dragoons
    Morgan's Rifle Corps
    Harrison's 1st Continental Artillery Regiment

Appendix C
Organization of the Virginia Continental Line
1777-78

Muhlenberg's Brigade
 1st Virginia
 5th Virginia
 9th Virginia
 13th Virginia

Weedon's Brigade
 2nd Virginia
 6th Virginia
 10th Virginia
 14th Virginia

Woodford's Brigade
 3rd Virginia
 7th Virginia
 11th Virginia
 15th Virginia

Scott's Brigade
 4th Virginia
 8th Virginia
 12th Virginia
 Grayson's Additional Battalion

Morgan's Rifle Corps

1st Continental Light Dragoons

1st Continental Artillery Regiment

Bibliography

Baxter, James, ed. *The British Invasion from the North: Digby's Journal of the Campaigns of Generals Carleton and Burgoyne from Canada, 1776-1777.* New York: De Capo Press, 1970.

Bearss, Edwin C. *The Battle of Sullivan's Island and the Capture of Fort Moultrie: A Documented Narrative and Troop Movement Maps.* U.S. Dept. of the Interior, 1968.

Bellas, Henry Hobart, ed. *Personal Recollections of Captain Enoch Anderson, an Officer of the Delaware Regiments in the Revolutionary War.* Wilmington: The Historical Society of Delaware, 1896.

Boatner, Mark M. *Encyclopedia of the American Revolution.* New York: D. McKay Co., 1966.

Bollinger, Lee C. ed. *The Events of My Life: An Autobiographical Sketch by John Marshall.* Ann Arbor, MI & Washington, D.C.: Clements Library, University of Michigan and Supreme Court Historical Society, 2001.

Boyle, Joseph Lee. *Writings from the Valley Forge Encampment of the Continental Army,* Vol. 1-2. Bowie: Heritage Books Inc., 2000-2002.

Brock, R. A. ed. "Orderly Book of Capt. George Stubblefield, March 11, 1776," *Miscellaneous Papers…in the Collections of the Virginia Historical Society.* Richmond, VA: Virginia Historical Society, 1887.

Brock, R.A. ed., "George Gilmer to Thomas Jefferson in Papers, Military and Political, 1775-1778 of George Gilmer, M.D. of Pen Park, Albemarle Co., VA," *Miscellaneous Papers 1672-1865 Now First Printed from the Manuscripts in the Virginia Historical Society.* Richmond, VA, 1937.

Brown, Lloyed and Howard Peckman, ed. *Revolutionary War Journals of Henry Dearborn, 1775-1783*. Chicago: The Caxton Club, 1939.

Burgoyne, John. *A State of the Expedition from Canada*. New York Times & Arno Press, 1969.

Campbell, Charles, ed. *The Bland Papers: Being a Selection from the Manuscripts of Colonel Theodorick Bland Jr. of Prince George County, Virginia*, Vol. 1. Petersburg: Edmund & Julian Ruffin, 1840.

Campbell, Charles, ed. *The Orderly Book of that Portion of the American Army Stationed at or Near Williamsburg...March 18th, 1776 to August 28th, 1776*. Richmond, VA, 1860.

Clark Stephen. *Following Their Footsteps: A Travel Guide & History of the 1775 Secret Expedition to Capture Quebec*. 2003.

Commager, Henry Steele and Richard B. Morris, eds. *The Spirit of 'Seventy-Six: The Story of the American Revolution as Told by Participants*. Edison, NJ: Castle Books, 2002.

Cresswell, Nicholas. *The Journal of Nicholas Cresswell*. New York: The Dial Press, 1924.

Dandridge, Danske. "Henry Bedinger to --- Findley," *Historic Shepherdstown*. Charlottesville, VA: Michie Co., 1910.

Davies, K.G. ed. *Documents of the American Revolution*, Vol. 3. Shannon: Irish University Press, 1972.

Dorman, John ed. "Peter Bruin Pension Application," *Virginia Revolutionary Pension Applications*, Vol. 12. Washington, D.C.: 1965.

Fischer, David Hackett. *Washington's Crossing*. Oxford University Press, 2004.

Graham, James. *The Life of General Daniel Morgan*. Bloomingburg, NY: Zebrowski Historical Services Publishing Company, 1993.

Greenwood, Isaac J. ed. *The Revolutionary Services of John Greenwood...* 1775-1783. New York, 1922.

Harris, Michael C. *Brandywine: A Military History of the Battle that Lost Philadelphia but Saved America, September 11, 1777*. California: Savas Beatie, 2014.

Heitman, Francis B. *Historical Register of Officers of the Continental Army During the War of the Revolution, April 1775 to December 1783*. Washington, D.C., 1914.

Hening, William W. ed. *The Statutes at Large Being a Collection of all the Laws of Virginia.* Vol. 9, Richmond: J. & G. Cochran, 1821.

Jackman, Sydney, ed. *With Burgoyne from Quebec: An Account of the Life at Quebec and of the Famous Battle at Saratoga.* Toronto: Macmillan of Canada, 1963.

Johnston, Henry P. *The Battle of Harlem Heights.* London: Macmillian,1897.

Johnson, Henry P. *The Campaign of 1776 around New York and Brooklyn.* Brooklyn: Long Island Historical Society, 1878.

Ketchum, Richard M. *Saratoga,: Turning Point of America's Revolutionary War*. NY: Holt & Co., 1997.
Lamb, Roger. *An Original and Authentic Journal of Occurrences During the Late American War from Its Commencement to 1783*. Dublin: Wilkinson & Courtney, 1809.

Lee, Henry. *The Revolutionary War Memoirs of General Henry Lee.* New York: Da Capo Press, 1998.

Lesser, Charles H. ed. *The Sinews of Independence: Monthly Strength Reports of the Continental Army.* University of Chicago Press, 1976.

Luzader, John. *Saratoga: A Military History of the Decisive Campaign of the American Revolution.* New York: Savas Beatie, 2008.

Marshall, John. *The Life of George Washington*, Vol. 2. Fredericksburg, VA: The citizens Guild of Washington's Boyhood Home, 1926.

Martin, David. *The Philadelphia Campaign, June 1777 – July 1778.* Da Capo Press, 1993.

Martin, Joseph Plum. *Private Yankee Doolittle: Being a Narrative of Some of the Adventures, Dangers and Sufferings of a Revolutionary Soldier.* Eastern Acorn Press, 1998.

Mays, David John, ed. *The Letters and Papers of Edmund Pendleton*, Vol. 1. Charlottesville: The University Press of Virginia, 1967.

McIlwaine, H.R. ed. "Proceedings of the Council of State of Virginia, February 12, 1777," *Journals of the Council of State of Virginia*, Vol. 1. Richmond, 1931.

McGuire, Thomas. *The Philadelphia Campaign: Brandywine and the Fall of Philadelphia,* Vol. 1. Stackpole Books, 2006.

McGuire, Thomas. *The Philadelphia Campaign: Germantown and the Roads to Valley Forge*, Vol. 2. Stackpole Books, 2007.

Moore Frank, ed. "New Jersey Gazette, January 28, 1778," *Diary of the American Revolution*, Vol. 2. New York: Charles Scribner, 1860.

Moway, Bruce. *September 11, 1777: Washington's Defeat at Brandywine Dooms Philadelphia*. PA: White Mane Books, 2002.

Muhlenberg, Henry A. *The Life of Major-General Peter Muhlenberg of the Revolutionary Army*. Philadelphia: Cary and Hart, 1849.

Pausch, George. *Journal of Captain Pausch, Chief of the Hanau Artillery During the Burgoyne Campaign*. Translated by William L. Stone, Albany, NY: Joel Munsell's Sons, 1886.

Reed, John. *Campaign to Valley Forge*. Pioneer Press, 1965.

Roberts, Kenneth. ed., *March to Quebec: Journals of the Members of Arnold's Expedition*. New York: Country Life Press, 1938.

Rodney, Caesar, ed. *The Diary of Captain Thomas Rodney, 1776-1777*. Wilmington: The Historical Society of Delaware, 1888.

Rogers, Horatio, ed. *Hadden's Journal and Orderly Book: A Journal Kept in Canada and Upon Burgoyne's Campaign in 1776 and 1777*. Boston: Gregg Press, 1972.

Ryan, Dennis P. *A Salute to Courage: The American Revolution as Seen Through Wartime Writings of Officers of the Continental Army and Navy*. NY: Columbia University Press, 1979.

Sandor, Gregory B., ed. *Journal of the Public Store at Williamsburg, 1775-1776*. Columbus, Ohio, 2015.

Sanchez-Saavedra, E.M., ed. *A Guide to Virginia Military Organization, 1774-1787*. Westminster, MD: Willow Bend Books, 1978.

Scheer, George and Hugh Rankin, eds. *Rebels and Redcoats: The American Revolution Through the Eyes of Those Who Fought and Lived It.* Da Capo Press, 1957.

Schnitzer, Eric. "Battling for the Saratoga Landscape," *Cultural Landscape Report: Saratoga Battle, Saratoga National Park,* Vol. 1. Boston, MA: Olmsted Center for Landscape Preservation.

Showman, Richard, ed. *The Papers of General Nathanael Greene,* Vol. 2. Chapel Hill: University of North Carolina Press, 1980.

Smith, Paul H. ed. *Letters of Delegates to Congress: 1774-1789,* Vol. 1. Washington, D.C.: Library of Congress, 1976.

Smith, Samuel. *The Battle of Brandywine.* Monmouth Beach, NJ: Philip Freneau Press, 1976.

Smith, Samuel, *The Battle of Princeton.* Monmouth Beach, NJ: Philip Freneau Press, 1967.

Stryker, William S. *The Battles of Trenton and* Princeton. Old Barracks Association, 2001.

Taylor, Robert J., ed. *The Papers of John Adams,* Vol. 3. Cambridge, MA: Belnap Press, 1789.

Thacher James, M.D. *Military Journal of the American Revolution.* Gansevoort, New York: Corner House Historical Publications, 1998.

Tustin, Joseph, trans. & ed. *Diary of the American War: A Hessian Journal.* New Haven: Yale University Press, 1979.

Tyler, Lyon, ed. *Tyler's Quarterly Historical and Genealogical Magazine,* Vol. 12. Richmond, VA: Richmond Press, Inc. 1931.

Ward, Harry M. *Duty, Honor, or Country: General George Weedon and the American Revolution*. Philadelphia: American Philosophical Society,1979.

Ward, Harry M. *Major General Adam Stephen and the Cause of American Liberty*. Charlottesville: Univ. Press of Virginia, 1989.

Wilkinson, James. *Memoirs of My Own Times*, Vol. 1. Philadelphia: Abraham Small, 1816.

Wright, Jr., Robert K. *The Continental Army*. Washington D.C.: Center of Military History United States Army, 1989.

Wrike, Peter. *The Governor's Island.* Gwynn, VA: The Gwynn's Island
  Museum, 1993.

----- *The Lee Papers*, Vol. 1-2. New York Historical Society, 1872.

Collections

American Archives

Force, Peter, ed. *American Archives, Fourth Series*, Vol. 2-6. Washington, DC: U.S. Congress, 1839-1843.

Force, Peter, ed., *American Archives, Fifth Series*, Vol. 1-3. Washington, DC: U.S. Congress, 1848-1853.

Journals of the Continental Congress

Ford, Worthington C., ed. *Journals of the Continental Congress, 1774-1789*, Vol. 2-9. Washington D.C.: U.S. Government Printing Office, 1905-1907.

Naval Documents of the American Revolution

Clark, William, ed. *Naval Documents of the American Revolution.* Vol. 1-4, Washington: 1964-1969.

Morgan, William J. ed. *Naval Documents of the American Revolution,* Vol. 5-6. Washington: 1970-1972.

The Papers of George Washington

Chase, Philander D., ed. *The Papers of George Washington, Revolutionary War Series,* Vol. 1-3. Charlotte: University Press of Virginia, 1985-1988.

Chase, Philander D. and Frank E. Grizzard, Jr., eds. *The Papers of George Washington, Revolutionary War Series,* Vol. 6. Charlottesville: University Press of Virginia, 1994.

Chase, Philander D. ed. *The Papers of George Washington, Revolutionary Series,* Vol. 7. Charlottesville: University Press of Virginia, 1997.

Grizzard Jr., Frank E. ed. *The Papers of George Washington, Revolutionary War Series,* Vol. 8. Charlottesville: University Press of Virginia, 1998.

Chase, Philander D., ed. *The Papers of George Washington, Revolutionary War Series,* Vol. 9. Charlottesville, University Press of Virginia, 1999.

Grizzard, Jr., Frank E., ed. *The Papers of George Washington, Revolutionary War Series,* Vol. 10. Charlottesville: University Press of Virginia, 2000.

Chase, Philander D. and Edward G. Lengel, eds. *The Papers of George Washington, Revolutionary War Series,* Vol. 11. Charlottesville: University of Virginia Press, 2001.

Grizzard, Jr., Frank E. and David R. Hoth, eds. *The Papers of George Washington, Revolutionary War Series,* Vol. 12. University Press of Virginia, 2002.

Revolutionary Virginia: The Road to Independence

Scribner, Robert L., and Brent Tarter, eds. *Revolutionary Virginia: The Road to Independence*, Vol. 3-6. Virginia Independence Bicentennial Commission, University Press of Virginia, 1978-81.

Tarter, Brent, ed. *Revolutionary Virginia: The Road to Independence*, Vol. 7 Part One. Virginia Independence Bicentennial Commission, University Press of Virginia, 1983.

Periodicals

Boyle, Joseph Lee, ed. "From Saratoga to Valley Forge: The Diary of Lt. Samuel Armstrong," *The Pennsylvania Magazine of History and Biography,* Vol. 121, No. 3. July 1997.

Dearborn, Henry. "A Narrative of the Saratoga Campaign – Major General Henry Dearborn, 1815," *The Bulletin of the Fort Ticonderoga Museum*, Vol. 1, No. 5. January, 1929.

Egle, Wm. H., ed. "The Journal of Captain William Hendricks and Captain John Chambers," *Pennsylvania Archives, 2nd Series*, Vol. 15. 890.

Flickinger, Floyd B. ed. "The Diary of Lieutenant William Heth while a Prisoner in Quebec, 1776," *Annual Papers of the Winchester Historical Society*, Vol. 1. 1931.

Katcher, Philip. "They Behaved Like Soldiers: The Third Virginia Regiment at Harlem Heights", *Virginia Cavalcade,* Vol. 26, No. 2. Autumn 1976.

Maxwell, William, ed. "A Recollection of the American Revolutionary War," *Virginia Historical Register and Literary Companion,* Vol. 6. (1853).

McMichael, James. "The Diary of Lt. James McMichael of the Pennsylvania Line, 1776-1778," *The Pennsylvania Magazine of History and Biography*, Vol. 16, no. 2. 1892.

Muhlenberg, Peter. "Orderly Book of Gen. John Peter Gabriel Muhlenberg, March 26-December 20, 1777," *The Pennsylvania Magazine of History and Biography*, Vol. 35. 1911.

Neville, Gabriel. "The B Team of 1777: Maxwell's Light Infantry," *Journal of the American Revolution.* (Online: April 10, 2018).

Peale, Charles Wilson. "Journal of Charles Wilson Peale," *Pennsylvania Magazine of History and Biography*, Vol. 38. Philadelphia: The Historical Society of Pennsylvania, 1914.

Sergeant R. "The Battle of Princeton," *The Pennsylvania Magazine of History and Biography,* Vol. 20, No. 1. 1896.

Sullivan, Thomas. "Before and After the Battle of Brandywine: Extracts from the Journal of Sergeant Thomas Sullivan of H.M. Forty-Ninth Regiment of Foot", *The Pennsylvania Magazine of History and Biography,* Vol. 31. Philadelphia: Historical Society of Pennsylvania, 1907.

Tarter, Brent, ed. "The Orderly Book of the Second Virginia Regiment: September 27, 1775 – April 15, 1776," *The Virginia Magazine of History and Biography*, Vol. 85, No. 2. April, 1977.

Uhlendorf Bernarld, and Edna Vosper, trans. and eds. "Letters of Major Baurmeister During the Philadelphia Campaign," *The Pennsylvania Magazine of History and Biography,* Vol. 59. Philadelphia: Historical Society of Pennsylvania, 1935.

Williams, William W., ed. "Robert Magaw to the Carlisle Committee of Correspondence, August 13, 1775," *Magazine of Western History*, Vol. 4. May-October, 1886.

—— "Orderly Book of General Muhlenberg, March 26 – Dec. 20, 1777," *Pennsylvania Magazine of History and Biography*, Vol. 35, No. 1. 1911.

Newspapers

Dixon and Hunter, "October 28, 1775," *Virginia Gazette*,

Dixon and Hunter, "January 31, 1777," *Virginia Gazette*.

Dixon and Hunter, "Dec. 26, 1777," *Virginia Gazette*.

Pinkney, "November 2, 1775," *Virginia Gazette*.

Purdie, "August 4, 1775," *Virginia Gazette*.

Purdie, "November 10, 1775," *Virginia Gazette*.

Purdie, "March 1, 1776," *Virginia Gazette*.

Purdie, "March 15, 1776," *Virginia Gazette Supplement*.

Purdie, "April 19, 1776," *Virginia Gazette*.

Purdie, "May 17, 1776," *Virginia Gazette*.

Purdie, "May, 26 1776," *Virginia Gazette*.

Purdie, "June 28, 1776," *Virginia Gazette*.
Purdie, "February 28, 1777," *Virginia Gazette*, Supplement.

New Jersey Gazette, "January 14, 1778".

Unpublished Works

Bacheller, Nathaniel, Letter, October 9, 1777, Copy on file at Saratoga National Historical Park.

Butler, Lt. Col. Richard, to Col. James Wilson, January 22, 1778," Gratz Collection, Case 4, Box 11, Historical Society of Pennsylvania.

Hill, Reverend William. "Sketch of Daniel Morgan: Notes and Sermon." Virginia Historical Society.

McDonald, Bob. transcribed, "Brigadier General George Weedon's Correspondence Account of the Battle of Brandywine, September 11, 1777," Manuscript is held by the Chicago Historical Society.

Posey, Thomas. *Revolutionary War Journal, Jan. 12, 1776 to May 24, 1777.* Thomas Posey Papers. Indiana Historical Society Library, Indianapolis, IN.

Posey, Thomas. *A Short Biography of the Life of Governor Thomas Posey.* Thomas Posey Papers. Indiana Historical Society Library, Indianapolis, IN.

Stephen, Adam. "General Adam Stephen to the Board of War, November 8, 1776," *Papers of the Continental Congress*, Vol. 1.

Index

1st Continental Light Dragoon Regiment, 172, 176-178
1st Virginia Regiment, 10, 12, 16, 18, 33, 43, 49-52, 56-60, 66, 73, 77, 84, 98-99, 106, 110, 130, 145, 149, 150, 153, 175, 272, 299, 301-305
1st Virginia State Garrison Regiment, 267-68
2nd Virginia Regiment, 10, 12, 17-18 28-29, 33, 37, 39, 41, 43, 45, 47, 49, 50-51, 53-54, 57-59, 77, 97-98, 108, 161, 175, 299, 301-305
3rd Virginia Regiment, 57, 60, 65, 68, 75, 77, 87, 91, 98-104, 106, 110-111, 116-117, 124-125, 130, 138, 145, 176, 184-186, 210-213, 217, 299, 301-305
4th Virginia Regiment, 61, 72, 98, 107, 130, 142, 152, 175-176 271, 299, 302-305
5th Virginia Regiment, 54, 57, 61-62, 73, 76, 130, 133, 156, 159, 169-170, 175, 189, 299, 302-305
6th Virginia Regiment, 50, 54, 56-57, 62, 77, 84, 87, 92, 106, 111, 116, 130, 140, 157, 169, 175, 270-272, 299, 302-305
7th Virginia Regiment, 54, 57, 63, 76-77, 84-85, 88-89, 91, 108, 158, 176, 189, 299, 301, 303-305
8th Virginia Regiment, 46, 54, 57, 59, 63-64, 75-82,

92-94, 107, 158, 176, 299, 303-305
9th Virginia Regiment, 46, 54, 64-65, 71,-72 108, 192, 203, 206, 267, 295, 299, 303-305
10th Virginia Regiment, 160, 176, 299, 303-305
11th Virginia Regiment, 161-162, 176, 179, 299, 303-305
12th Virginia Regiment, 162-163, 176, 300, 303-305
13th Virginia Regiment, 163, 175, 293, 300, 303-305
14th Virginia Regiment, 164, 176, 293, 300, 303-305
15th Virginia Regiment, 164, 176, 300, 303-305
14th Regiment, British, 28, 36, 42, 49
40th British Regiment, 147, 201, 262-263
49th British Regiment, 192, 203, 206
55th British Regiment, 146-147, 152
60th British Regiment, 64

2nd Virginia Convention, 9
3rd Virginia Convention, 10, 12, 60
4th Virginia Convention, 45, 47, 72
5th Virginia Convention, 83, 156

A

Abercromby, Robert, 238
Accomack County, VA, 65
Adams, John, 7, 188,
Albemarle County, VA, 17, 63, 65, 164,
Alexander, Morgan, 17, 59
Alexander, William (Lord Stirling), 110-111, 116-117, 123-124, 129-130, 167-168, 184, 199, 209-210, 213, 215
Alexandria, VA, 57, 173
Amelia County, VA, 16, 59, 164
Amelia District, 12, 16, 58
Amherst County, VA, 17, 62, 161
Anburey, Thomas, 237-239, 254
Anderson, Enoch, 123
Anderson, Richard Clough, 62

Arbuckle, Matthew, 162
Armstrong, Samuel, 229
Arnold, Benedict, 8, 21-22, 25, 27, 48, 226, 246, 249-251
Arnold's March to Quebec, 21-27
Arundel, Dohicky, 71-72, 89, 155
Ashby, Stephen, 163
Augusta County, VA, 17, 63-64, 66, 161-162
Aylett, William, 14

B
Balcarress, Colonel, 297
Balcarress, Lt. Col., 297
Ball, Burgess, 74
Ballard, Robert, 19, 68, 330
Baurmeister, Carl, 240, 246, 249, 250, 251, 386
Baylor, George, 194, 218
Beale, Robert, 163, 167, 172, 175, 187
Beall, Isaac, 61
Bedford County, VA, 17, 62, 164
Bedinger, Henry, 4, 6, 113-115

Bemis Heights, NY, 227, 240, 243-244
Bennington, Battle of, 225
Berkeley County, Virginia, 4, 17-18, 59, 61, 64, 127, 163
Berry, Thomas, 64
Birmingham Meeting House, PA, 210-212
Blackwell, William, 161
Bland, Theodorick, 156, 168, 171-172, 176-178, 289-290
Bordentown, NJ, 139
Botetourt County, VA, 17, 63, 163
Bound Brook, NJ, 171
Bowman, Abraham, 64
Bowyer, Michael, 162
Bowyer, Thomas, 163
Brandywine, battle of, 199-217
Brent, John, 61
Breymann, Colonel, 243, 249-252
Bruin, Peter, 5, 162,
Brunswick County, VA, 17, 61, 164,
Buckingham County, VA, 19, 59, 62
Buckingham District, 17

Buckner, Mordecai, 62, 70, 169
Buford, Abraham, 31-32,164
Bullett, Thomas, 12-13, 70
Burgoyne, Gen. John, 187, 223, 225-228, 230, 235, 238, 240-245, 247-249, 252-256, 270
Burlington, NJ, 129, 144, 277
Burton Point, VA, 84
Butler, Richard, 231-232, 246, 251, 278, 287

C

Cabell, Samuel Jordan, 62
Cadwalader, Gen. John, 129, 132, 140, 150-151
Cambridge, MA, 81
Campbell, Richard, 64, 92-93,
Campbell, William, 17-18, 58
Caroline County, VA, 10, 58-59, 62, 161
Caroline District, 17
Carrington, Edward, 72, 155
Casey, Benjamin, 163

Charles City County, VA, 17, 62
Charleston, battle of (1776), 80-83
Charleston, SC, 79-80, 83, 92-94, 187
Charlotte County, VA, 16, 61, 164
Cherokee Indians, 98, 166
Chester, NY, 185
Chester, PA, 107, 217, 286, 288
Chesterfield County, VA, 16, 62, 64, 164
Chew, Benjamin, 262
Chilton, Capt. John, 60, 99-103, 105, 117-118, 124, 184-186
Christian, William, 12, 57, 73, 98
Claiborne, Buller, 59
Clark, Jonathan, 64
Clark, William, 147, 150
Clinton, Gen. Henry, 80-81, 240,
Clinton, George, 104, 224, 295
Cocke, Nathaniel, 63
Continental Congress, 3, 4, 11, 42, 46-47, 50, 56, 71, 84, 97, 107-108, 165,

Cooches Bridge, battle of, 190-194
Cooches Bridge, DE, 190, 192-193
Cornwallis, Gen. Charles, 78, 80, 114, 140-142, 144, 153, 278, 289
Crawford, William, 61, 108, 189
Cresswell, Nicholas, 160
Crockett, Joseph, 63
Croghan, William, 64
Cropper, John, 65
Culpeper County, VA, 60, 64, 160-161, 164
Culpeper District, 16, 57
Culpeper Minute Battalion, 31, 33-34, 49
Cumberland County, VA, 12, 16, 63, 161

D

Daingerfield, William, 76, 87
Darke, William, 64
Davies, William, 16, 58,
Davis, Thomas, 65
Dearborn, Henry, 226, 229, 232, 238-239, 242-243, 245, 247, 251, 253
Denny, Samuel, 72

Dickenson, Edmund, 58
Dickinson, Capt., 274
Digby, William, 229, 233, 236, 253
Dilworth, PA, 210, 216, 218
Dinwiddie County, VA, 17, 62, 164
Dumfries, VA, 54, 57, 173
Dunmore County, VA, 64, 163
Dunmore, John Murray, Earl of, 9, 18, 28, 30, 33, 35-36, 38, 42, 49-51, 75, 77, 84-87, 89-91, 94, 98
Durkee, John, 130

E

Eastern Shore, VA, 10, 45-46, 54, 57, 64, 77, 108, 158
Elizabeth City County, VA, 17
Elizabeth City District, 17, 59
Elizabethtown, NJ, 117
Elliot, Thomas, 62
Enos, Roger, 25
Epps, Francis, 12, 57, 99
Essex County, VA, 16, 63

Ewald, Johann, 171, 191-193, 211-212, 280, 289
Ewing, Gen. James, 129, 132

F

Fairfax County, VA, 17, 60, 62, 77, 161
Faulkner, Ralph, 62
Fauntleroy, Henry, 62
Fauquier County, VA, 16, 60, 105, 162-162
Febiger, Christian, 161
Ferguson, Patrick, 201
Fermoy, General, 172, 173
Fermoy, General de, 123, 140-141
Fincastle County, VA,12, 17, 63-64, 163-164
Fitzgerald, John, 60
Fleming, Charles, 63
Fleming, John, 16, 58, 150, 153
Fleming, Thomas, 64
Flora, Billy, 39
Flying Camp, 97, 99
Fontaine, William, 17-18, 59
Fordyce, Capt. Charles, 40-41

Fort Lee, 109, 112, 116-117, 155
Fort Mercer, NJ, 270, 274-278
Fort Mifflin, 270-271, 275, 304
Fort Pitt, PA, 175
Fort Stanwix, 276
Fort Stanwix, NY, 225
Fort Ticonderoga, NY, 252
Fort Washington, 109, 111-117, 155,161, 302
Fort Western, Maine, 21-22, 25
Franklin, James, 161
Fraser, Simon, 228, 230, 247-248
Frederick County, Virginia, 4, 64, 161-162
Frederick, MD, 5-6
Fredericksburg, VA, 60
French and Indian War, 1, 10, 12, 51, 60-61, 107, 166

G

Gallihue, Charles, 162
Gaskins, Thomas, 62
Gates, Gen. Horatio, 127, 225-228, 240-242, 244-245, 252, 254-255, 267

Georgia, 92-93, 107, 158
Germain, Lord, 86, 238, 281
Germantown, battle of, 259-266
Gibson, George, 17-18, 58, 66-69, 268
Gibson, John, 163
Gilchrist, George, 65
Gillison, John, 161
Gist, Nathaniel, 166
Gloucester County, VA, 116, 54, 56, 63, 65, 76, 77, 84, 87, 91, 108
Gloucester Courthouse, VA, 57
Gloucester District, 58
Glover, John, 277
Goochland County, VA, 16, 65, 164
Grant, Peter, 166
Grayson, William, 165
Grayson's Additional Battalion, 165, 174, 176, 300, 303-305
Great Bridge, VA, 35, 75,
Great Bridge, battle of, 35-44, 51, 57, 160, 301
Green, John, 16, 18, 33, 57, 99, 270

Greene, Christopher, 275, 277
Greene, Gen. Nathanael, 112, 129, 132, 134, 136, 145, 216, 259, 261, 263-265, 276-279
Greenwood, John, 131-133, 137
Gregory, William, 62
Grey, Gen. Charles, 289
Griffith, David, 104, 125
Gwynn's Island, battle of, 84-91

H
Hadden, James, 232-233
Hager, George, 282
Halifax County, VA, 16, 63, 164
Halifax, NC, 78
Hamilton, Gen. James, 228, 232, 235-236
Hampshire County, VA, 18, 64, 163
Hampton, battle of, 28-33
Hand, Edward, 140-142, 145, 152
Hanover County, VA, 10, 62, 164
Harlem Heights, battle of, 102-106

Harrison, Charles, 72, 155, 304
Harrison, Robert, 205-206, 208
Hawes, Samuel, 59
Hayes, John, 65
Hazen, Moses, 175
Head of Elk, MD, 188
Helphinstine, Peter, 64
Henrico County, VA, 16, 58, 61-62
Henrico District, 16
Henry, John Joseph, 27, 30, 31
Henry, John Joseph, 23, 25
Henry, Patrick, 9-10, 12-16 33, 50-52. 57, 60
Hessians, 114, 129, 133-134, 136, 138-139, 160, 205
Heth, William, 161 190, 200, 202, 205
Hill, Reverend William, 254-255
HMS *King Fisher*, 33-34
HMS *Mercury*, 28
HMS *Otter*, 28-29
Homer, Christian, 155
Hopkins, Samuel, 62
Howe, Col. Robert, 42, 49, 62, 69

Howe, Gen. William, 68-69, 102, 111, 116-117, 123, 126-127, 140, 153, 181, 183-188, 191-192, 194, 196-197, 200-202, 206, 210-211, 214, 218-219, 221, 241, 262, 264, 270, 275, 279-281, 283-284, 290
Huntington, Gen. Jedediah, 277
Hutchings, Thomas, 62

I
Innes, James, 72-73, 118, 164
Intolerable Acts, 3
Iron Hill, DE, 190, 192-193
Isle of Wight, Virginia, 16, 61, 164

J
Jagers (German riflemen), 171, 178, 191-193, 212, 280, 282
James City County, VA, 17
Jamison, John, 156
Jefferson, Thomas, 31, 34
Johnson, James, 62
Johnson, Jr., Thomas, 60

Johnson, William, 162
Johnston, George, 17, 59
Jones, John, 62
Jones, Llewellin, 156
Jones, Strother, 166
Jones, Wood, 59
Jouett, Matthew, 63
Joynes, Levin, 65

K
Kelly, William, 170
Kennett Square, PA, 197, 200, 202
King and Queen County, VA, 16, 63, 162
King George County, VA, 17, 60, 101
King William County, VA, 14, 16, 62-63, 164
Knowlton, Thomas, 102, 104-106
Knox, Henry, 136, 262
Knox. James, 64
Knyphausen, General, 201, 203-204, 206-207

L
La Fayette, Marquis de, 278,
Laird, David, 161
Lamb, Roger, 233, 249

Lancaster County, VA, 17, 59, 62
Lancaster District, 17
Lancaster, PA, 220
Langdon, Jonathan, 163
Laurens, John, 282
Lawson, Robert, 61
Learned, General, 245, 250
Lee, Gen. Charles, 70-80, 82-83, 87-88, 92-93, 117, 119, 127,
Lee, Light Horse Harry, 157, 200, 286-291
Lee, Richard Henry, 74
Leitch, Andrew, 60, 98, 104-106
Leslie, Capt. Samuel, 38-40
Leslie, Peter, 41
Lewis, Charles, 164
Lewis, Gen. Andrew, 51, 77, 86-89, 9, 174-175
Lexington and Concord, Battle of, 1, 3, 9
Lindsay, Lieutenant, 286-287, 289-290
Loudoun County, VA, 17, 60, 62, 162, 165
Louisa County, VA, 16, 60, 164

Lunenburg County, Va, 16, 62, 164
Lynn, George, 30
Lynn, Lt. William, 66, 68-69

M
Madison, Rowland, 163
Magaw, Robert, 112-113, 115
Markham, John, 12, 16, 57
Marshall, John, 60, 120162, 207
Marshall, Thomas, 60, 98, 210-213
Martin, Joseph Plum, 264
Maryland rifle companies, 8, 48, 108-109, 161
Mason, David, 164
Mason, Nathaniel, 61
Mathews, George, 64, 265
Mathews, Thomas, 61
Mawhood, Charles, 146-148, 152
Maxwell, William, 189-191, 193195, 200-201, 204-209, 269
McClanahan, Alexander, 63, 91
McDougall, Alexander, 172, 260, 264

McGuire, James, 166
McKee, William, 162
McMichael, James, 148
McWilliams, William, 60
Meade, Richard Kidder, 17, 41, 59, 164
Mechlenburg County, VA, 62
Mercer, Hugh, 60, 65-69, 75-76, 87, 97, 99, 111, 123, 125, 129-130, 145-151, 153, 175
Middlesex County, VA, 16, 63
Mifflin, Thomas, 106
Mitchell, Joseph, 163
Monroe, Lt. James, 138-139
Montgomery, Gen. Richard, 48
Montresor, John, 191-192, 214
Moore, Cleon, 166
Morgan, Daniel, 4-8, 17, 21-23, 25, 27-28, 48, 59, 161, 179-182, 189, 202, 224-232, 234, 239-240, 242, 244-246, 249, 252 254-256, 276-277, 280-283, 286, 288

Morgan's Rifle Corps, 178-184
Morris, Joseph, 231, 251, 283
Morrison, George, 23, 26
Morristown, NJ, 153, 167, 171
Morton, John, 61
Moulder, William, 150
Moultrie, William, 81-83
Mountjoy, John, 161
Moylan, Stephen, 176
Muhlenberg, Peter, 63, 80, 82-83, 92-94, 174-175, 199, 217, 260-261, 264, 267, 269, 276, 285, 293-295, 305
Musgrave, Thomas, 262

N
Nansemond, County, VA, 16, 61, 164
Nash, Francis, 199
Nelson Jr., Thomas, 58
Nelson, John, 157
Nelson, William, 63
Neville, John, 162
New Brunswick, NJ, 116-118, 153
New Kent County, VA, 17, 62

Newcastle, VA, 12
Nicholas, George, 17, 29-30, 32-33, 59, 159-160
Nicholas, Robert Carter, 29
Norfolk, VA, 16, 28, 33, 35-36, 41-42, 49, 54, 57-58, 75, 77, 87, 164
Northumberland County, VA, 17, 62, 164

O
Ohio County, VA, 162-163
Orange County, VA, 16, 62-63, 161

P
Page, John, 31, 34, 110
Parker, Commodore Peter, 81-83,
Parker, Josiah, 61, 169,
Parker, Richard, 17, 59, 190, 200
Parramore, Thomas, 65
Patterson, Thomas, 62
Patton, John, 176
Pausch, George, 235-236,
Peachy, William, 61, 76
Pendleton, Edmund, 34, 51, 53, 66
Percy, Hugh, 114

Philadelphia, PA, 3-4, 11, 50, 75, 93, 126, 146, 150-151, 167, 173, 185-188, 196-197, 199, 218-219, 221, 259, 265-267, 269-270, 275, 277, 279, 284, 286, 288, 290-291
Piscataway, NJ, 182
Pittsylvania County, VA, 17, 58, 62, 164
Pittsylvania District, 17
Pleasants, John, 62
Poor, Gen. Enoch, 232, 245, 247
Porterfield, Charles, 162, 200, 202, 205, 207
Posey, Thomas, 63, 84-85, 89-91, 182-183, 246
Powell, Levin, 165
Prince Edward County, VA, 16, 61, 163
Prince George County, VA, 17, 59, 62, 164
Prince William County, VA, 12, 17, 56, 59-60, 77, 162, 165
Prince William District, 17
Princess Anne County, VA, 16, 28, 35, 58, 75, 77, 164
Princess Anne District, 16

Princeton, battle of, 145-153
Princeton, NJ, 120, 137, 139-140, 144-147, 150, 152, 155, 159, 167, 175,

Q
Quebec, Canada, 8, 21, 28, 48, 161, 226

R
Radnor Meeting House, PA, 286-288
Raleigh Tavern, 52
Rall, Col. Johann, 127, 137-139
Rawlings, Moses, 109, 113-114, 117
Read, Isaac, 61, 98
Reed, Col. Joseph, 156, 157
Reed, Joseph, 128
Reuber, John, 115
Richeson, Holt, 63, 164
Richmond County, VA, 17, 54, 56-57, 61-62, 164
Richmond, VA, 9,
Ridley, Thomas, 61
Riesdel, Gen. von, 227
Rodney, Thomas, 150,
Ross, David, 165

Roxbury, MA, 7, 48
Ruffin, Thomas, 62
Russell, Andrew, 62
Russell, William, 16

S
Saratoga, battle of, 227-252
Savannah, GA, 93
Sayres, John, 16, 58
Scammell, Alexander, 294
Schuyler, Philip, 21, 189, 223-225
Scott, Charles, 12, 37, 46, 58, 60, 98, 142-143, 169-170, 174, 176, 190, 199, 210, 213-215, 260, 285
Scruggs, Gross, 62
Senter, Isaac, Dr., 23, 26
Sergeant R, 147, 149, 151-152
Shawnee Indians, 51
Sheldon, Elisha, 177
Shelton, Clough, 161
Shepherdstown, VA, 4
Simms, Charles, 163, 190, 200, 203
Slaughter, George, 64
Smallpox, 87, 173

Smallwood, Col. William, 126
Smallwood, Hebard, 166
Smith, Arthur, 41
Smith, Granville, 166
Smith, Gregory, 63
Smith, Samuel, 271-274
Smith, William, 162
Snead, Thomas, 65
Southampton County, VA, 17, 59, 61
Southampton District, 17
Spencer, Joseph, 63
Spotswood, Alexander, 12, 39, 58, 108
Spotswood, John, 161
Spotsylvania County, VA, 12, 17, 60, 62, 63, 160-161
Squire, Capt., Matthew, 28-30, 32
St. Augustine, FL, 28, 91-92
Stafford County, VA, 17, 60, 161
Stephen, Adam, 61, 107-108, 111, 116-117, 123, 125, 129-131, 133-134, 140, 142, 146, 152, 199, 209-210, 214-215, 263, 269

Stephenson, David, 64
Stephenson, Hugh, 4, 6, 8, 21, 48-49, 108-109, 113, 299, 301
Stevenson, John, 64
Stevens, Edward, 160
Stevens, Richard, 161
Stirling, Lord (Gen. William Alexander), 110-111, 116117, 123-124, 129-130, 167-168, 184, 199, 209-210, 213, 215
Stocking, Abner, 27
Stubblefield, George, 62
Suffolk, VA, 35, 49, 54, 56-57, 59, 75-77
Sullivan, Gen. John, 158, 161, 163, 165, 168
Sullivan, Gen. John, 129, 131, 133-134, 137, 145-147, 151-152, 168, 177, 181, 199, 209-210, 213-215, 259, 261-263, 265, 269
Sullivan, Thomas, 192-193, 203-206, 208
Surry County, VA, 17, 61, 164
Sussex County, VA, 17, 61, 164
Symmes, John, 161

T
Talaiferro, William, 17
Temple, Benjamin, 156, 172
Terrill, Henry, 62
Thacher, James, 6
Thornton, John, 60
Tomkies, Charles, 63
Towles, Oliver, 62
Trenton, battle of, 128-139
Trenton, 2nd battle of, 140-144
Tripplett, Thomas, 166

V
Valley Forge, PA, 284-296
Varnum, Gen. James, 277, 285,
Virginia Committee of Safety, 12, 14, 29, 31, 33-34, 37, 42, 50-51, 53-54, 56, 59, 72, 75, 78-79
Virginia-Maryland Rifle Corps, 108

W
Waggener, Andrew, 162
Walker, Thomas, 65
Wallace, Gustavus Brown, 60, 101,
Warren, James,
Warwick County, VA, 17

Washington, Gen. George, 4, 7-8, 21, 48, 71, 73, 84, 87, 94, 97, 99, 102, 106, 109, 111-112, 116-117, 119-120, 123-134, 136-140, 142-146, 151-153, 156-159, 165-167, 169, 171-174, 176-178, 180-181, 183-184, 186-190, 194-196, 199-200, 205, 209, 216-221, 223-225, 241, 255-256, 259-262, 265-272, 274-279, 281-287, 290-292, 295
Washington, John, 61
Washington, Samuel, 126
Washington, William, 60, 138-139
Watchung Mountains, 153
Watkins, Jr., John, 61
Wayne, Anthony, 199, 259, 263-264, 269
Webb, John, 63
Weedon, George, 60, 98-99, 106, 110, 124, 174-175, 190, 199, 210, 212, 216-217, 260, 276, 285, 293-295
Wemys, James, 201
West Augusta, VA, 17, 64, 163

West, Charles, 60
Westfall, Abek, 64
Westmoreland County, VA, 17, 98, 164
Wetherall, Frederick, 282
White Plains, NY, 109-111, 117-118, 301
White, Joseph, 138
Whitemarsh, battle of, 275-276, 279-284
Wiederhold, Andreas, 114
Wilkerson, James, 161
Wilkinson, James, 131, 138, 141-144, 147, 152, 228, 230-232, 234, 244, -245, 247-250
William and Mary, College of, 12
Williamsburg, 12, 13
Williamsburg, VA, 9-12, 14, 16-18, 21, 28-29, 31, 33-34, 36, 45, 50, 52-54, 56-58, 65, 68-71, 73, 75-77, 83-84, 86-87, 91-92, 97-98, 108, 157, 161, 164
Willis, Francis, 166
Willis, John, 166
Willis, Lewis, 160
Wilmington, DE, 107, 188
Wilmington, NC, 78

Winchester, VA, 4-5, 247
Wood, James, 162
Woodford, Col. William,
   10, 12, 16-17, 28, 31, 33,
   35, 37-39, 42, 49, 51, 56,
   58, 65, 77, 98, 107-108
Woodford, Gen. William,
   174, 176, 190, 199, 210,
   213-215, 217, 260, 263,
   285, 293, 295, 305
Woodson, Samuel, 65
Woodson, Tarleton, 161
Yates, Bartholomew, 149, 153
York County, VA, 17, 58

www.ingramcontent.com/pod-product-compliance
Lightning Source LLC
Chambersburg PA
CBHW050836230426
43667CB00012B/2020